Spinoff 1999

National Aeronautics and Space Administration
Office of Aero-Space Technology
Commercial Technology Division

Developed by
Publications and Graphics Department
NASA Center for AeroSpace Information (CASI)

Foreword

Nearly a century ago, the Wright brothers set in motion the age of flight. Just 30 years ago, humans landed on the Moon. Today, NASA launches a new space mission every 11 weeks. With the technologies developed to go to the Moon, Mars, and beyond, we also have improved the quality of life here on Earth.

The generation that will lead us into the new millennium will build the highway of the future. The highway of the future will send astronauts beyond Earth's orbit safely, take business ventures to space cheaply, and help America explore the cosmos confidently.

The interplanetary highway, like today's highway, will have destinations. It will reach laboratories to harvest new technologies in biomed, biotech, robotics, advanced materials and sensors, to name a few. Its on-ramps will be open to everyone, not just NASA astronauts and Russian cosmonauts. It will even have its own communication system, an interplanetary internet. It will go well beyond Earth's orbit and the Moon, and reach other planetary bodies, asteroids, comets, and other solar systems. It may even help us answer the question: Are we alone? Along the way it will also bring new technologies we haven't even dreamed about yet.

We almost take for granted the returns on NASA's past investments: global communications, TV satellite broadcasts, extended weather forecasting, digital imaging, fire retardant materials, smoke detectors, computer barcoding, disposable diapers, the pacemaker, scratch resistant glasses, cordless power tools, remote monitoring devices for intensive care patients, and countless contributions to commercial aircraft engines and air traffic systems.

One person who certainly does not take NASA's work for granted is a young Alabama boy. He has a rare genetic disorder that prevents him from going into the daylight without experiencing nerve inflammation, severe blistering, and other serious effects. Recently, he was able to play outside during daylight, for the first time, wearing a protective suit developed from the same technology that protects our astronauts from ultraviolet radiation in space.

This is what NASA is about—opening up the space frontier, solving the mysteries of the universe, and bringing the discoveries into our homes. *Spinoff 1999* is a tribute to the spirit of the men and women who dare to dream everyday about building the highway of the future.

Daniel S. Goldin
Administrator
National Aeronautics and Space Administration

Introduction

In this year 1999, we find ourselves on the threshold of a new century with its exciting discoveries to come and challenges to meet. Clearly in the last decades of this century, some of our most significant achievements in industry and our lifestyle have been brought about by new technology. We can expect this trend to continue and perhaps accelerate.

Because of the great scientific and technical challenges of NASA missions, the demands of our programs result in many new technological discoveries. Meeting the NASA aeronautical and space goals has necessitated cutting-edge technical advances across a broad spectrum that embraces virtually every scientific and technological discipline.

As in previous years, we are proud to present *Spinoff 1999* as one demonstration of the successful transfer of NASA technology, resulting in products and services that benefit you, members of your family, and your business or industry.

Research and development in such fields as advanced sensors, new materials, enhanced digital imaging techniques, advanced power systems—to name a few—have generated technology for decades. That technology is available to U.S. commercial industry through a vast electronic marketplace, easily accessible via NASA TechTracS on the NASA Commercial Technology Internet site **<http://www.nctn.hq.nasa.gov>**.

Multiple uses of NASA technology have never been easier and more in line with industry needs and practices. U.S. industry, by working in partnership with NASA during the development phase of new technology, can speed the application of that technology to new products and services, thereby reducing time to market and public availability. Moreover, dual-use of NASA technology offers a less costly, complementary means of bringing new products to market.

More than 1,200 Spinoff products and processes have emerged from the multiple uses of technology needed for NASA missions. Each has contributed some measure of benefit to the national economy, productivity, or quality of life—some with moderate contributions, but many with benefits of significant order and considerable economic value.

This NASA mission to share the wealth of our technology with the public sector is accomplished through the Commercial Technology program. It is our aim to broadly apply technical knowledge. Consequently, the vast storehouse of NASA technology is a national resource bank available for commercial "spinoff" applications.

This publication is a yearly report to the public, which documents successful outcomes of our program. It is intended to enhance the awareness of the technology that was used by NASA and business, and of the potential for public benefit. As the past successes have been significant, we look forward to an even brighter and more fruitful result. We invite you to share in these recent results.

Spinoff 1999 is organized into three sections: 1) **Aerospace Research and Development** highlights major research and development efforts currently carried out at the 10 NASA field centers, which could result in future commercial products; 2) **Technology Transfer and Commercialization** features this year's center highlight, NASA's John H. Glenn Research Center at Lewis Field and its commercialization efforts, as well as the mechanisms in place nationwide to assist U.S. industry in obtaining, transferring, and applying NASA technology, expertise, and assistance; 3) **Commercial Benefits – Spinoffs** describes recent successes in industry in the development of commercially available products and services resulting from the transfer of NASA technology.

Dr. Robert L. Norwood
Director, Commercial Technology Division
National Aeronautics and Space Administration

Spinoff developments highlighted in this publication are based on information provided by secondary users of aerospace technology, individuals, and manufacturing concerns who acknowledge that aerospace technology contributed wholly or in part to development of the product or process described. Publication herein does not constitute NASA endorsement of the product or process, nor confirmation of manufacturers' performance claims related to the particular spinoff development.

Contents

Foreword

Introduction

Aerospace Research and Development
NASA Headquarters and Centers

    NASA Headquarters 10

    Ames Research Center 14

    Dryden Flight Research Center 16

    Glenn Research Center 18

    Goddard Space Flight Center 20

    Johnson Space Flight Center 22

    Kennedy Space Center 24

    Langley Research Center 26

    Marshall Space Flight Center 28

    Stennis Space Center 30

    and the Jet Propulsion Laboratory 32

Technology Transfer and Commercialization

    Center Spotlight 36

    Network /Affiliations 40

Commercial Benefits—Spinoffs

    Health and Medicine 46

    Transportation 48

    Consumer/Home/Recreation 54

    Environment and Resources Management 58

    Computer Technology 66

    Industrial Productivity/Manufacturing Technology 76

NASA Success and Education—A Special Feature 91

NASA Commercial Technology Network 96

Aerospace Research and Development

NASA's Strategic Plan encompasses leadership, direction, and purpose, guiding all NASA field centers to follow four Strategic Enterprises*: Aero-Space Technology, Earth Science, Human Exploration and Development of Space, and Space Science.

The research and development efforts inherent in these four enterprises will ensure continued success in scientific and technical advances beneficial to NASA and the scientific and engineering community of the United States. NASA organizations are involved in bold, versatile, and unique endeavors, which are demonstrated on the following pages, highlighting current pioneering and breakthrough efforts at each field center.

These efforts will contribute to the continued preeminence for the United States, not only in matters of science and technology, but also in the transfer of technology to private industry. This will result in commercially developed products and services, facilitating a continued global leadership position for our economy, as well as benefits for humankind in general.

*For more details about the Strategic Enterprises go to <http://www.nasa.gov/enterprises.html>.

NASA Headquarters

NASA Headquarters and Centers

If past is prelude, the National Aeronautics and Space Administration (NASA) has placed the United States on an enviable trajectory into the 21st century. This pathway builds upon four decades of innovative research, cutting-edge technological development, and an unending quest to push back the frontiers of aeronautical and space flight.

Established in 1958, NASA has evolved into an amalgam of four Strategic Enterprises: Human Exploration and Development of Space, Space Science, Earth Science, and Aero-Space Technology. NASA Headquarters is the manager for the agency's multi-pronged exploratory mission of scientific research, investigating the solar system and beyond, and technology development and transfer. The space agency's goals stretch out some 25 years. There are many components to this roadmap into the 21st century.

Human Outpost

NASA's Human Exploration and Development of Space (HEDS) Enterprise includes the International Space Station, Space Shuttle, and Life and Microgravity research. HEDS seeks to bring the frontier of space fully within the sphere of human activity for research, commerce, and exploration.

The Space Station is the largest peacetime scientific and technological project in history. Thousands of individuals in 16 countries are part of a formidable challenge to build the Space Station. When fully operational in 2004, this orbiting complex will have the pressurized volume of laboratory space equivalent to two jumbo jet airliners.

"In five years, the Space Station will be complete and serving as an outpost for humans to develop, use, and explore the space frontier. The Space Station will greatly expand research opportunities, leading to exploration breakthroughs, scientific discoveries, technology development, and new space products," explains NASA Administrator Daniel Goldin.

Research onboard will help thwart diseases, such as cancer, diabetes, and AIDS. By unmasking the effects of gravity, the basic properties of materials can be studied, perhaps yielding products of direct benefit to Earth. The Space Station can serve as a testbed, indeed the platform, to evaluate equipment and flight-qualify humans for long stints in microgravity, in order to regain a foothold on the Moon and place the first footprints on Mars.

The HEDS Enterprise is committed to join with the private sector to spark opportunities for commercial development in near-Earth space as a key to future settlement. HEDS activities include use of resources, even those provided by comets and asteroids, to sustain a human presence beyond Earth. Safe, reliable, low-cost space transportation is critical to the goals of the HEDS Enterprise. NASA's Space Shuttle program, operating as

An image of the International Space Station taken by an STS-96 crewmember during a fly-around by Discovery.

the world's most versatile launch system, is pledged to flight safety as a top priority. Meeting the shuttle manifest of flights, improving system supportability and reliability, and reducing cost follow in that order of priority.

HEDS is also implementing a shuttle upgrade program to improve reliability, performance, and longevity of Space Shuttle operations to meet Space Station requirements.

Looking for Life

What is the destiny of the human species? Are we alone in the universe? How did the universe, galaxies, stars, and planets form and evolve? NASA has started on the path to answer these and other fundamental questions.

The Chandra x-ray observatory rests inside the payload bay of the orbiter Columbia.

Four basic themes have been identified for the Space Science Enterprise: Sun-Earth connection; exploration of the solar system; structure and evolution of the universe; and the astronomical search for origins. In addition, the origins and distribution of life in the universe is a fifth theme, which cuts across the other four.

NASA's Space Science Enterprise has become a shining example of a new way of thinking at the agency. In the last few years, billions of dollars from planned spending have been removed. The philosophy of "faster, better, cheaper" has been embraced, with the size, complexity, and cost of spacecraft missions having been reduced significantly, while at the same time, increasing or maintaining their scientific capability.

Extensive robotic investigation of Mars is now underway or on the books. Plans have been drafted for the first return samples from Mars to arrive on Earth in 2008. Sets of Mars orbiters and landers will transform the mysterious red planet into a more familiar world. Robotic spacecraft will probe Martian weather and climate and help piece together the intriguing story of whether Mars has been, or is currently, an abode for life. Ultimately, these explorations will set the stage for the first footfalls on Mars early in the 21st century, as humankind extends outward from our home planet.

Mars is not the only frontier for exploration. The Lunar Prospector has identified what are apparently caches of ice hidden within craters at the Moon's north and south poles, resources protected from the Sun's warming rays. The Galileo spacecraft has produced striking images of Jupiter's moon, Europa. Scientists believe an ocean may exist underneath Europa's icy facade, perhaps an ocean teeming with life.

Closing in on its target, the Near-Earth Asteroid Rendezvous (NEAR) will begin a yearlong investigation of the asteroid Eros. The Cassini spacecraft is on a 7-year journey to study Saturn, its moon, and its rings. Onboard is the European Space Agency-built Huygens probe that will parachute in 2004 onto Saturn's enigmatic moon, Titan. The Stardust spacecraft is now trekking toward comet Wild-2, on a mission to return a sample of cometary dust to Earth in 2006. Similarly, the Comet Nucleus Tour (CONTOUR) will image and spectrally map at least three very different comets and the dust flowing from those objects between 2002 and 2008.

Orbiting observatories, like the 12.5-ton Hubble Space Telescope, have taken stunning images of the surrounding cosmos. Joining Hubble is Chandra, outfitted to address fundamental questions in science by obtaining x-ray images of neutron stars, black hole candidates, quasars, and active galaxies. Also to be launched is the Space Infrared Telescope Facility (SIRTF) with duties to scrutinize young galaxies in formation, study quasars, and probe the very nature of black holes.

But to investigate the way our universe formed following the Big Bang demands a new type of spaceborne observatory. Working with industry teams, NASA's Next-Generation Space Telescope (NGST) is to harness state-of-the-art technologies, allowing scientists to peel back time and observe an era when stars and galaxies started to form.

NASA Headquarters *Continued*

NASA Headquarters and Centers

Little stirs the soul of exploration more than the question: Are there other habitable planets like Earth circling nearby stars? A systematic approach to answer this query involves the Space Interferometry Mission (SIM), the Terrestrial Planet Finder (TPF), and the Life Finder observatory. NASA is positioning itself to directly detect Earth-sized planets around stars within 100 light years of Earth. If these planets are detected, spaceborne instruments should be able to pick up the signs of whether or not these worlds are conducive to life. The ultimate goal is to take a picture with a resolution high enough to see oceans, mountain ranges, cloud cover, and the continents of an Earth-like world.

Mission to Planet Earth

NASA's Earth Science Enterprise has been organized to better understand the entire Earth system and the effects of natural and human-induced changes on the global environment. To this end, the field of Earth System Science is being pioneered. An emerging interdisciplinary field, this research considers Earth's land surface, oceans, atmosphere, ice sheets, and life as both dynamic and highly interactive.

The Earth Science Enterprise is strategically structured to study five major Earth System Science areas: land-cover and land-use change; seasonal-to-interannual climate variability and prediction; natural hazards research and applications; long-term natural climate variability and change research; and atmospheric ozone research.

To comprehend our changing planet, new knowledge and tools for better weather forecasting, urban and land-use planning, agriculture, and other areas that yield both economic and environmental benefit are being sought. Space, air, and ground-based platforms are providing the scientific foundation for policies that strive for sustainable development of Earth.

Astronaut James H. Newman holds one of the hand rails on the Unity connecting module during the early stages of a 7-hour, 21-minute spacewalk.

The space agency's Earth Observing System (EOS) era has begun with the recent launch of the Landsat 7, to be followed by the EOS-AM-1. This scientific duo, and others being readied for orbit, will collect needed data to help answer key questions about Earth. With EOS, how land and coastal regions are changing over time can be evaluated. Also, forecasting precipitation a year in advance may be feasible. Furthermore, by determining the probabilities of floods and droughts, predicting changes in Earth's climate a decade to a century in advance will be a reasonable expectation. Lastly, monitoring ozone depletion to determine the effectiveness of efforts to control harmful chemicals is planned. EOS launches will continue through the second decade of the 21st century.

A series of lightweight, low-cost science missions tagged Earth System Science Pathfinders (ESSP) are also being readied. The first two ESSP missions are scripted, one designed to make the first global inventory of the world's forests, and the other focused on measuring the variability of Earth's gravity field. As part of NASA's New Millennium program to validate cutting-edge technology, an Earth Orbiter-1 mission will demonstrate an advanced land imaging system with multispectral capability starting in 1999. Another technology validation mission is the Space-Readiness Coherent Lidar Experiment to be flown on a Space Shuttle in 2001. This experiment will test whether a space-based sensor can precisely measure atmospheric winds from the Earth's surface to a height of 10 miles.

NASA's goals are to predict the weather, climate, and natural disasters with a much higher accuracy and to make forecasts on a seasonal to interannual basis. "If we can predict, we can prepare, maybe even prevent," Goldin believes. "Hopefully, within 25 years, we'll be able to make multi-decade predictions of climate and environment, so we can better manage our resources for sustainable development...globally, regionally, and locally," he adds.

Pillars of Progress

The Aero-Space Technology Enterprise is responsible for answering the question: How does the space agency enable revolutionary technological advances that provide air and space travel for anyone, anytime, anywhere more safely, more affordably, and with less impact on the environment, while improving business opportunities and global security? NASA's vision of future flight runs from supersonic travel to highways in the sky and huge flying wings.

NASA's agenda in this area is focused on "Three Pillars" for success: Global Civil Aviation; Revolutionary Technology Leaps; and Access to Space.

The X-38 Crew Return Vehicle is released from high altitude, so the project team can conduct aerodynamic verification maneuvers, among other tests.

Initiatives have been put in place to work on aviation system technologies that support pilots and air traffic controllers. NASA has teamed with the Federal Aviation Administration (FAA) to prioritize technology efforts that can ensure aviation safety is maximized for the flying public. NASA's Aviation Systems Capacity (ASC) Program is one such activity, looking at modernizing and improving the Air Traffic Management System and the introduction of new vehicle classes that can potentially reduce congestion.

A new effort has begun, called the Ultra-Efficient Engine Technology Program, established to enable the next breakthroughs in propulsion systems that could spawn a new generation of high-performance, operationally efficient, economically viable, and environmentally compatible U.S. aircraft.

Small planes also mean big business. A partnership between government and industry is revitalizing the U.S. light airplane industry. The challenge is to create a small aircraft transportation system as an alternative to short-range automotive trips for both private and business transportation. NASA is joining forces with the FAA and U.S. industry to make personal aircraft as affordable as luxury automobiles. Making flying as safe as driving on the interstate, and increasing the ease with which pilots learn and maintain their flying skills are challenges meant to help invigorate the general aviation industry.

Space transportation technology work at NASA translates into the goal of affordable and reliable access to space. Toward this end, the space agency has partnered with U.S. industry and the Department of Defense to build next-generation reusable space transportation.

The challenge is to lower the high cost of access to space. At present, the expense of placing payloads into low-Earth orbit is roughly $10,000 a pound. This expensive price tag cripples the hope for a dynamic, creative, and productive U.S. space enterprise. A major NASA focus is reducing within 10 years the payload cost to low-Earth orbit by an order of magnitude, from $10,000 to $1,000 per pound. By the year 2020, that cost to low-Earth orbit is to drop from the $1,000s to the $100s per pound.

Reusable launch vehicles (RLVs) require advancements in several areas. These include propulsion, composite fuel tanks and structures, improved thermal protection systems, and diagnostic sensors. Rapid turnaround and frequent flights of RLVs using small ground crews are also among the major objectives.

Two unpiloted RLVs—the X-33 and X-34—are being readied for maiden flights. The wingless, wedge-shaped X-33 flight demonstrator is a suborbital prototype for a single-stage-to-orbit vehicle. Featuring vertical takeoff and glider-like landings, the X-33 will approach speeds of 15 times the speed of sound at altitudes of 50 miles. The smaller, winged, air-launched X-34 also lands horizontally. This vehicle will approach speeds 8 times the speed of sound at 50 miles altitude.

NASA has also embarked on the Future-X series of demonstrations. To validate technologies beyond the X-33 and X-34, Future-X sets the stage for creating a new generation of space launchers, to be built faster and cheaper than previous vehicles. Through Future-X, NASA can readily test and validate new, state-of-the-art space transportation technologies in flight.

"NASA remains committed to providing the American taxpayer with the best possible space and aeronautics program in the world. Our accomplishments demonstrate we are capable of that. We are determined to continue that tradition. I truly believe the best is yet to come," Goldin concludes. ❖

Ames Research Center
NASA Headquarters and Centers

America's aerospace proficiency depends on the ability to research, develop, and transfer cutting-edge aeronautical technologies. To this end, there is need for unique integration of computation, simulation, ground and flight experimentation, and information sciences.

This integration is a tenet of the Ames Research Center in Mountain View, California. Other missions of Ames include answering fundamental questions concerning evolution, such as astronomical and planetary environments, the adaptation of living systems to space, and the health of our planet. Additionally, Ames is dedicated to designing, developing, and delivering integrated information systems technologies and applications to further advance technologies for space flight.

In late 1998, Ames officials announced a visionary concept for 21st century research, development, and education. A new Ames Research Complex will encompass 2,000 acres of federal property owned by NASA, with the objectives of establishing a world-class, shared use, research and development campus in conjunction with local communities, and involving partnerships with government, academia, private industry, and non-profit organizations.

"To become part of our development, any partnership must further the NASA mission and enhance life in America in the 21st century," notes Henry McDonald, Ames director. "We believe that by establishing these partnerships we will also strengthen the technological leadership of Silicon Valley and enhance the well-being of our communities," he adds.

Ames' aeronautics work is also leading NASA's research and technology development efforts for rotorcraft and powered-lift aircraft.

Illustrating NASA's aeronautics and computational prowess is Ames' Center-TRACON Automation System (CTAS). This set of tools is designed to help air traffic controllers manage the increasingly complex air traffic flows at airports. CTAS benefits air traffic controllers by reducing stress and workload, thereby benefiting air travelers by reducing delays and increasing safety.

At the heart of CTAS is a new approach to air traffic control, expressed as human-centered automation. Conceived and prototyped at Ames, CTAS is based on algorithms, software, and human interfaces, fashioned to help predict aircraft trajectories with high, real-time accuracy to schedule aircraft to land at runways with the least possible delay and to provide methods for communicating between CTAS and controllers through special graphical interfaces. These innovations, fused with radar tracks and weather data, are melded in a complex real-time software system that generates CTAS information.

The VMS is controlled by large digital computers that are programmed to represent the aircraft and the external environment.

Ames' aeronautics work is also leading NASA's research and technology development efforts for rotorcraft and powered-lift aircraft. Likewise, center researchers play a lead role in creating world-class flight simulators. Ames' computer scientists are even delving into new "smart" software that will enable aviators to control and safely land disabled airplanes. This type of intelligent flight control system employs experimental "neural network" software developed at Ames.

Ames is also working on technology known as Vertical Motion Simulation (VMS). With its unique motion capability, VMS represents one of the world's most technically advanced research and development flight simulators. It is used to simulate a complete spectrum of flight vehicles, including the Space Shuttle, civil and military transports, and rotary-wing aircraft.

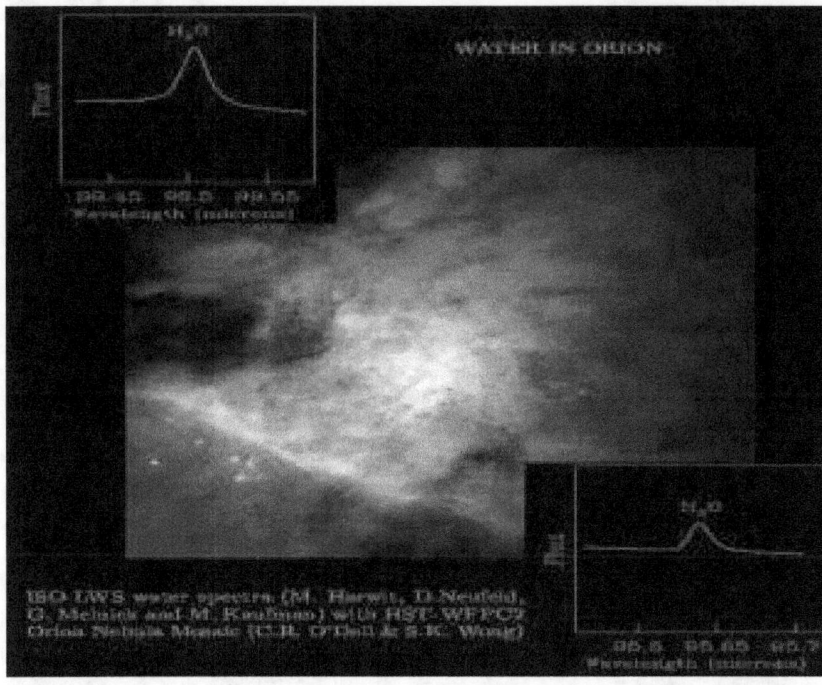

Ames' research in astrobiology involves searching for sources of water in the galaxy, including this study of the Orion nebula.

How did life begin? Are there other habitable planets? These are fundamental scientific queries. To begin chipping away at these questions in search of answers, Ames is leading NASA's astrobiology program. Astrobiology is an emerging interdisciplinary field that deals with life in the universe: its origin, evolution, distribution, and destiny. Research tasks involve the formation and evolution of habitable worlds, life's evolution and the biochemical cycles in the Earth's oceans and atmosphere, and the potential for biological evolution beyond an organism's planet of origin.

A key aspect of biological research is the use of a major facility—the Gravitational Biology Laboratory—on the International Space Station. Ames' scientists and engineers are responsible for developing this on-orbit facility. Once a part of the Space Station, the Gravitational Biology Laboratory will provide multiple habitats to support a variety of organisms, a large centrifuge with a selectable rotation rate to simulate a range of gravity levels, a holding rack to house specimens at microgravity, and a fully equipped workstation/glovebox for the research crew. The laboratory also includes microscopes, freezers, and other hardware to conduct experiments.

Ames served as mission control for the highly successful Lunar Prospector spacecraft. Circling the Moon since January 1998, the probe mapped the Moon's magnetic and gravity fields, charted the resources of that celestial body, and studied concentrations of hydrogen found at the north and south lunar poles. Mission scientists interpret the hydrogen concentrations to be vast deposits of frozen water—a valuable resource for future human explorers of Earth's neighbor.

In the fast-paced world of computers, Ames has teamed up with an industry-leading computer company to build the world's first working parallel supercomputer. The computer, called "Steger," named after Joseph Steger, a deceased Ames computer scientist, links 256 microprocessors to create one huge supercomputer. Using Steger, scientists are attempting to work out the largest aeronautical problems that NASA has tried to solve by using calculations 10 times larger than those done before. The parallel supercomputer approach has the potential to cut months from the design cycle of a new airplane, as one example of its attributes.

Yet another Ames-led computer innovation is making the slogan "the network is the computer" a truism for more and more applications. Toward this end, the center is playing a leading role in the Next-Generation Internet (NGI) project. NGI research will develop networks that are 100 to 1,000 times faster than today's Internet.

The future of space exploration begins with advancements in computer technology here on Earth. ❖

Dryden Flight Research Center
NASA Headquarters and Centers

Sky high successes are part of the rich tradition of the Hugh L. Dryden Flight Research Center, situated at Edwards, California.

Projects at Dryden over the last 50 years have led to major advancements in the design and capabilities of many civilian and military aircraft. Dryden engineers played a vital role in breaking the "sound barrier" with the X-1 aircraft in 1947, a historical first in aeronautics. Just as in the past, this NASA research center is flying the newest breed of vehicles today, built to break new ground while flying over clear desert skies.

Among a roster of current projects, is Dryden's Advanced Control Technology for Integrated Vehicles (ACTIVE) program. Outstanding flying qualities of a highly modified F-15 aircraft have been achieved when thrust vectoring of engine exhaust is integrated into the flight control system. The same aircraft also recently served as a test bed to evaluate a new neural network-based intelligent flight control software, designed to allow a pilot to safely control a damaged aircraft. Additional research activities at Dryden include the F-18 Active Aerolastic Wing that improves maneuverability and aircraft performance, and management of flight tests of Russia's Tu-144LL aircraft, a cooperative venture between the United States and Russia in high-speed research.

NASA's Environmental Research Aircraft and Sensor Technology (ERAST) program at Dryden seeks to demonstrate that an aircraft can fly at the edge of the atmosphere, at altitudes up to 100,000 feet, for hours on end. Centurion is a prototype of an innovative, remotely piloted, solar-powered airplane being developed under the ERAST effort. At this ultra-high altitude, Centurion—also known as the "Helios prototype"—can serve as an upper-atmosphere research vehicle, hauling sensors and instruments to record important data for scientists some 19 miles below. Flight endurance of the Centurion is about 14 to 15 hours.

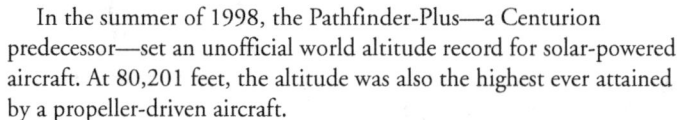

In the summer of 1998, the Pathfinder-Plus—a Centurion predecessor—set an unofficial world altitude record for solar-powered aircraft. At 80,201 feet, the altitude was also the highest ever attained by a propeller-driven aircraft.

Pathfinder-Plus and the Centurion are providing a step-by-step approach to ERAST's ultimate "eternal airplane." This vehicle, named the Helios, will attempt to fly early next century for up to four months at a time at altitudes from 50,000 to 70,000 feet, while hauling a hefty 200-pound payload of science gear.

NASA's two ER-2 Airborne Science aircraft are based at Dryden. These flying laboratories collect information about our surroundings, including Earth resources, celestial observations, atmospheric chemistry and dynamics, and oceanic processes. ER-2 aircraft are also used for electronic sensor research and development, satellite calibration, and satellite data validation. These planes have supported ozone depletion study campaigns over Antarctica, the Arctic, Chile, and other areas to gauge ozone loss. The ER-2 has played a significant role in charting the extent of fires that raged through Yellowstone National Park.

Once an astronaut is in orbit and there is a problem, how can he or she get home in a hurry? Answering that question is crucial in the upcoming era of the International Space Station. Dryden engineers and NASA's Johnson Space Center have begun flight-testing the X-38. This prototype spacecraft could become the first new human space-

Centurion is a remotely piloted prototype of a solar-powered airplane that can fly to high altitude, remaining aloft for extended periods. High above Earth, the NASA-sponsored research plane will carry science equipment to perform upper-atmosphere measurements.

The X-38 prototype Crew Return Vehicle (CRV) is airborne under the wing of NASA's B-52 aircraft. Engineers are testing the X-38 as a step toward vehicles that would return crew members of the International Space Station in emergency situations.

craft built in the past two decades that travels to and from orbit. The vehicle is being developed at a fraction of the cost of past human space vehicles. The goal is to take a "cheaper, better, faster" approach to spacecraft construction, taking advantage of available equipment and already-developed technology for as much as 80 percent of the spacecraft's design.

The goal of the innovative X-38 project is to develop a Crew Return Vehicle, a "lifeboat," for the Space Station. As the Space Station grows, holding as many as seven crew members, two X-38-derived Crew Return Vehicles may be attached to the orbiting facility. The first test drops of the X-38 have taken place over Dryden.

Dryden is the site for a rigorous flight research program, dubbed the "Hyper-X," or X-43A. This program is set to demonstrate airframe-integrated, air-breathing engine technologies. Such an engine concept promises to increase payload capacity for future vehicles, including reusable space launchers and aircraft that fly more than five times the speed of sound. Orbital Sciences Corporation's Pegasus launch vehicle is being modified to push the X-43A to high speeds. In the near future, the first of a series of X-43A flights are scheduled, with the Hyper-X craft geared to reach 10 times the speed of sound. These hypersonic speeds will demonstrate ramjet/scramjet engine technologies. If successful, the X-43A aircraft is designed to fly faster than any previous air-breathing aircraft.

Regarding the purpose of flight research, the work at Dryden is perhaps best characterized by the late Dr. Hugh L. Dryden himself, "...to separate the real from the imagined, and to make known the overlooked and the unexpected problems...." ❖

Hyper-X program's X-43A vehicle is to be powered by air-breathing scramjet engine technologies, pushing the craft to 10 times the speed of sound.

Glenn Research Center

NASA Headquarters and Centers

The dance of a flame in the microgravity of space, reducing the roar of aircraft engines, and spearheading novel concepts that may make interstellar travel a reality one day are among the research fields being pursued at the NASA Glenn Research Center at Lewis Field in Cleveland, Ohio.

In March 1999, the Lewis Research Center was officially renamed the NASA John H. Glenn Research Center at Lewis Field, in recognition of America's first astronaut to orbit the Earth and his four terms of senatorial work on behalf of Ohio and the nation in the U.S. Congress. George William Lewis was the research director for the National Advisory Committee for Aeronautics, the predecessor to NASA.

"The blending of names reflects the pioneering research in aerospace technology that employees have performed throughout the center's history and will continue to perform in the future," observes Glenn's director, Donald J. Campbell.

Power, propulsion, and communications technologies are high-priority research domains being advanced at the center, so that U.S. leadership in these areas is ensured. Since the early formative years of the space agency, work in turbomachinery at Glenn has been underway.

Glenn's turbomachinery research goals are focused on technology that enables aircraft engine manufacturers to design advanced turbine engines. This research has included the development of new high-temperature materials, coatings, and lubrication systems; development and verification of design, modeling, and computational codes for workstation computers and visualization tools; basic research in flow physics and heat transfer; and combustion research. All of this work has been supported by materials testing, component testing, and prototype testing in facilities that simulate operating conditions of turbomachinery.

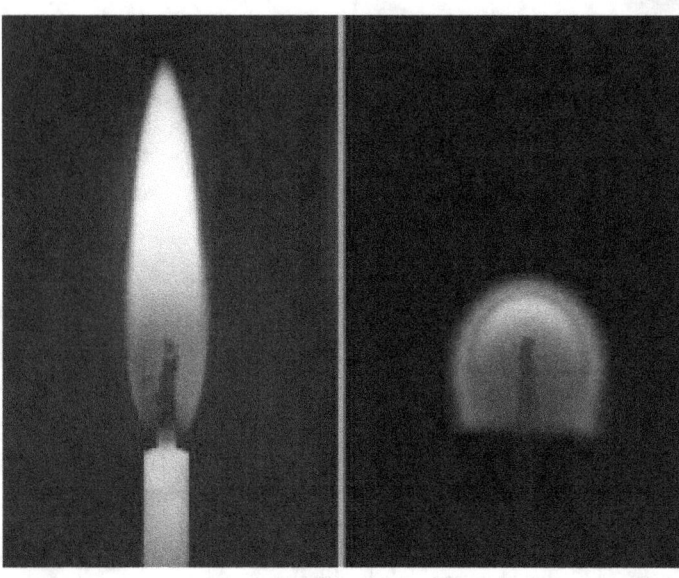

The effect of microgravity on combustion processes has been a major area of research guided by Glenn Research Center scientists. Knowledge about the physics of combustion may lead to more efficient combustion of materials, and therefore a savings of fuel.

Along with the U.S. aeropropulsion industry, Glenn is enabling technology for the next generation of subsonic gas turbine engines, which are environmentally friendly, fuel efficient, durable, and globally-marketable. For example, Glenn engineers are taking the technological steps to reduce nitrous oxide emissions—prime contributors to the development of smog—from gas turbine engines. Experimental and analytical work is also being performed in three areas of engine noise reduction: active noise control for fans, advanced low-noise fan designs, and jet noise.

In the field of general aviation, Glenn's advanced subsonic technology work is striving to improve the safety, performance, and ease-of-use of general aviation aircraft. Efforts at the center include advanced propulsion sensors and controls to allow simplified intermittent combustion engine and aircraft control rather than the approach taken over the last 40 years. Gains are being made in civil tilt rotor aircraft, too. While this subsonic aircraft can take off and land using less

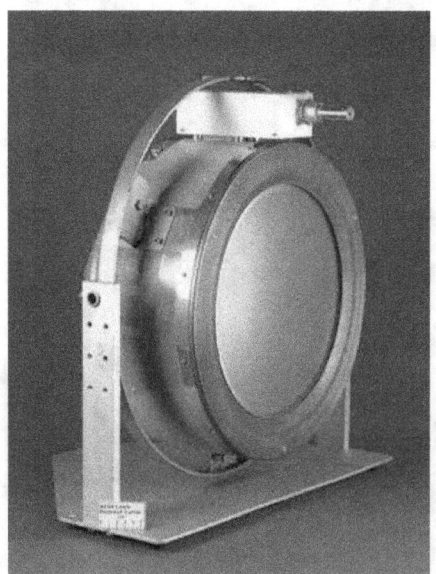

Work in ion propulsion technology at Glenn Research Center made possible the technology validation flight of the Deep Space 1 spacecraft. The fuel-efficient ion engine was tested in space in 1998-1999, and is to be utilized on 21st century deep space missions.

runway and appears viable in relieving air traffic congestion, Glenn experts are tackling noise, terminal-area operations, safety, weight reduction, and reliability issues.

High above Earth, the Advanced Communications Technology Satellite (ACTS) is trailblazing advanced communication technologies and services. Released into orbit by a Space Shuttle crew in 1993, the ACTS symbolizes Glenn's long history of developing state-of-the-art communications technology. The center has formed partnerships with industry, universities, and other government agencies to utilize the Ka-band spectrum through ACTS, to validate use of all-digital, high-bandwidth, on-demand, integrated multimedia services.

A major milestone in propulsion technology is the in-space shakeout of the ion engine aboard NASA's Deep Space 1, launched in late 1998. Glenn engineers have long been working on ion engines, technology that produces an almost imperceptible thrust that is equivalent to the pressure exerted by a sheet of paper held in the palm of a hand. While the xenon-fueled ion engine is slow to pick up speed, over the long haul it can deliver 10 times as much thrust per pound of fuel as liquid or solid fuel rockets. Ion drives are likely to be used on future deep space and Earth-orbiting missions that would otherwise be impractical or unaffordable with conventional propulsion systems.

Glenn's Fluids and Combustion Facility for the International Space Station is a modular, multi-user facility accommodating microgravity science experiments. In the absence of gravity, combustion takes place in a very different manner than on Earth. Gravity plays a role in why flames behave the way they do, smoke rises, and how large air circulation currents are established. These effects can mask the physical processes of combustion. Understanding what exactly is happening in the physics of combustion is sure to have major payoffs. Combustion powers our automobiles, generates our electricity, heats our homes, even cooks our food on the backyard grill, and adds many pollutants to our skies. Microgravity research in this area aboard the Space Station could offer ways of increasing combustion efficiency, thus reducing pollution, global warming, and production and transportation costs.

Taking a long view into the 21st century, current work at Glenn is studying futuristic propulsion concepts. Ultimate breakthroughs in space transportation are under review at the center involving propelling a vehicle without propellant mass, attaining the maximum transit speeds physically possible, and creating new energy production methods to power such devices. Nobody can predict where insights into these areas may take future robotic and human explorers. ❖

An array of microphones inside the Aero-Acoustic Propulsion Laboratory at the Glenn Research Center measures simulated aircraft engine noise during takeoffs. Acoustic data collected is used to certify aircraft engines, as specified by the Federal Aviation Administration (FAA).

Goddard Space Flight Center

NASA Headquarters and Centers

Goddard scientists are attempting to put the pieces of a global puzzle together to develop a conceptual Earth system model.

Chances are that the latest spectacular, headline-stealing discovery in space, physics, or astronomy is connected to research underway at the Goddard Space Flight Center. This center was established in 1959, becoming NASA's first major scientific laboratory dedicated entirely to space exploration.

Goddard is situated in Greenbelt, Maryland, taking on duties in space science, earth science, and technology development. The center is involved in implementing suborbital programs as well, using aircraft, balloons, and sounding rockets. This function is located at the Wallops Flight Facility on Wallops Island, Virginia.

The ongoing goal of NASA's Earth Science Enterprise is to integrate data to better study the Earth as a system. There are basic questions at the heart of such an investigation. What makes the Earth habitable? Why here, and on no other planet in our solar system do we find highly diversified life? One obvious reason is that Earth's climate system constitutes a thermostat unique in our solar system.

But the Earth is a dynamic system, with interrelated elements that feed into complex layers of hydrosphere, physical earth, and atmosphere, along with the biosphere. Goddard scientists are attempting to put the pieces of a global puzzle together to develop a conceptual Earth system model, but those components are highly interactive. NASA's Earth Observing System (EOS)—a series of science satellites—offers a way to decipher and scrutinize all the key interactions of the Earth's climate system. Some 25 launches to place 30 satellites into low Earth orbits by 2003 are now on the books.

Soon to join the recently orbited Landsat 7, are other major Earth science missions, including Terra, the first EOS morning satellite (AM-1), along with the first EOS afternoon satellite (PM-1), the Ice, Cloud and Land Elevation Satellite (ICEsat-1), and the atmospheric chemistry mission (CHEM-1).

Goddard scientists have already begun to watch the Earth's climate unfold. Launched in 1997, the Sea-viewing Wide Field-of-view Sensor (SeaWiFS) has yielded new insights into the impact of El Niño-related events, specifically on ocean life, such as phytoplankton blooms. SeaWiFS also eyed floods in China, Gobi and Sahara Desert dust storms, fires in Mexico, and the spiraling fury of hurricanes Bonnie and Danielle.

Also recently lofted was the Tropical Rainfall Measuring Mission (TRMM). This jointly U.S.- and Japanese-built satellite is measuring tropical and subtropical rainfall, lightning variability and distribution, and solar absorbed and Earth emitted radiation. TRMM is intended to enhance our knowledge of the vertical distribution of heating in

The purpose of TRACE is to study the Sun's magnetic fields in connection with the heating of its corona.

the atmosphere and improve basic understanding of worldwide atmospheric circulation.

While taking the pulse of Earth is crucial for personal and planetary well-being, a sense of wonderment about the surrounding universe also occupies many of Goddard's scientists and engineers. Goddard is readying for a 2000 launch of the Microwave Anisotropy Probe (MAP), which will be assigned the task of measuring temperature variations in microwave radiation from the cosmological Big Bang. A technology program is underway to fashion Constellation-X, a mission to study powerful objects and events, like supermassive black holes and exploding stars. Goddard is active in the Gamma Ray Large Area Space Telescope (GLAST), to be launched in 2005 to observe the most violent events in the universe with unprecedented detail.

The physical processes that link the Sun and the Earth are being evaluated by Goddard spacecraft. First results from the Goddard-managed Advanced Composition Explorer (ACE) have challenged the current understanding of the acceleration of particles caused by explosions on the Sun. Similarly, the Transition Region and Coronal Explorer (TRACE) relayed stunning observations of activity in the solar atmosphere. Powerful events on the Sun, such as flares, can trigger communication and power system failures on Earth.

The quest to peer even deeper into the universe is under the direction of Goddard through its Next-Generation Space Telescope (NGST) project. NGST represents challenges on numerous engineering fronts, from light-weight structures to multi-segmented, deployable mirrors. The mega-powerful NGST could be launched in 2007, built to see objects 400 times fainter than those currently studied with larger ground-based infrared telescopes or other spaceborne counterparts. Moreover, NGST must study objects with the image sharpness achieved by the Hubble Space Telescope.

Landsat 7 represents an unprecedented program to check the health of Earth.

From the immensity of the universe down to the size of a nano-satellite, Goddard engineers have begun a technology development effort to build disc-shaped spacecraft, weighing less than 22 pounds. A mere 12 inches wide, and just 4 inches thick, nano-satellites are viewed as a method to revolutionize the scientific investigations of key physical processes, in both the space science and Earth science arenas.

Goddard planners envision tens to hundreds of nano-satellites dispatched in space, flying in formation in order to make simultaneous measurements of an area of the sky. This approach would provide scientists more precise data than a single satellite taking a single measurement. For the first time, simultaneous measurements in both space and time will be resolved.

Whether developing missions that rewrite the text books to better understand our own planet or unlock the mysteries of the universe, Goddard is positioned to enter the 21st century on the cutting edge. ❖

Johnson Space Center
NASA Headquarters and Centers

Important pathways to the future are embodied in a variety of Johnson projects.

NASA's Center of Excellence for human operations in space is the Johnson Space Center in Houston, Texas. Since the 1960s, Johnson's Mission Control Center has been key to the success of every piloted U.S. space mission. From the early Gemini, Apollo, and Skylab projects to today's Space Shuttle flights, Johnson is the very hub of human space exploration.

Today, Johnson is NASA's lead center for the Space Shuttle program, the International Space Station program, space operations management, biomedical research and countermeasures, and the advanced human support technology program. The center's agency-wide assignments include extravehicular activity (space walks), robotics technology associated with human activities, space medicine, technology utilization on the Space Station, and long-range exploration mission planning and design.

Johnson is responsible for astronaut selection and training. About 150 men and women, diverse in heritage and background, form the nation's astronaut corps. The "Class of '98" consisted of 8 pilot and 17 mission specialist candidates, including school teacher Barbara Morgan, who was named as an Educator Mission Specialist. Of the 25 class members, 21 are male and 4 are female.

World-class facilities at Johnson are in use to prepare astronauts for building, then occupying, the Space Station. On December 4, 1998, Space Shuttle Endeavour lifted the U.S.-built Unity module into space. Shuttle astronauts successfully coupled Unity with a previously launched Russian-built module, Zarya. Thus began the step toward years of assembly needed to bring the Space Station into full operational use.

Important pathways to the future are embodied in a variety of Johnson projects. A major piece of the multi-nation Space Station program is under development at Johnson. The X-38 is an innovative, prototype spacecraft that could lead to the design, development, and construction of a emergency Crew Return Vehicle (CRV) for the Space Station. Drop testing of the Johnson-constructed X-38 has begun, with the CRV to be released from a Space Shuttle in the future, followed by descent through the Earth's atmosphere and a controlled landing. On-going testing of the TransHab inflatable module is also being conducted, with an eye toward possible use of this structure on the Space Station, or even as habitats on the Moon and Mars.

Continuing to draw scientific and public interest is the Johnson-led research into a 4.5-billion-year-old Martian meteorite. As reported in 1996, Johnson researchers observed what appears to be evidence of ancient primitive life on Mars. Further evidence to possibly support the claim has been provided by NASA, the Department of Energy, and university research teams. That evidence centers on finding microorganisms here on Earth similar in size and morphology to those in the Martian meteorite. This find, however, does not prove definitively that the features in the meteorite—known as ALH 84001—are biological in origin.

Johnson outreach into the academic community is strong, exemplified by the National Space Biomedical Research Institute. A partnership has been struck among Johnson, the Baylor College of Medicine, and six other universities. Joint research is centered on development and

John H. Glenn simulates rappelling from a troubled Space Shuttle

TransHab, the large-volume inflatable space vehicle, is a new design for a habitation element for lengthy space missions.

implementation of countermeasures to enable long-duration human space flight.

A business technology incubator couples Johnson with the University of Houston's (UH) science and business acumen. This UH/NASA Technology Commercialization Incubator is designed to help local small and mid-sized businesses commercialize space technology. NASA-patented technology can be licensed by entrepreneurs who, with the help of the business incubator, can bring to market new commercial products. UH is analyzing the commercial market, helping in business plan development and scouting out venture capital. Without such a helping hand, many start-up companies could not succeed.

Highlighting the value of moving NASA technology into the private sector is the use of spacesuit technology to combat porphyria. This genetic disorder causes extreme and potentially dangerous sunlight sensitivity that can lead to chronic skin disturbances. Johnson spacesuit experts, medical officials, and private industry teamed to devise a garment that protected a child from the sun's ultraviolet rays and other light sources.

Johnson also "floated" an innovative concept for stimulating college space research. Forty-seven teams of undergraduates from around the United States flew aboard NASA's KC-135A aircraft in a roller coaster-like flight profile over the Gulf of Mexico. The teams experienced repeated periods of about 25 seconds of floating in microgravity, enough time to conduct pre-selected sets of student-designed experiments.

In 1998, Johnson became the first NASA center and the first sizeable U.S. government organization to earn ISO 9001 certification. High marks were given to Johnson by third-party auditors who reviewed the center's management commitment, design control, documentation, purchasing, test and inspection, and corrective action procedures. ISO 9001 comprises the most detailed, comprehensive set of standard requirements for quality programs as established by the International Standards Organization.

Johnson director George W. S. Abbey points with pride at the workforce within and outside the gates of his sprawling complex of buildings. "The dedicated and innovative contributions of our federal and contractor workforce have enabled the human space flight program and the Johnson Space Center to excel," he notes. "Their accomplishments exemplify the spirit of exploration that will be needed to meet the challenges that lie before us as we approach the 21st century." ❖

Kennedy Space Center
NASA Headquarters and Centers

Handling complex space launchers and satellites demands sharp skills and the ability to integrate and test.

The John F. Kennedy Space Center in Florida is the historic departure point for human space explorers, situated on Florida's central Atlantic "space coast". As each returning Shuttle mission breaks the sound barrier before touching down at the center's landing strip, it's obvious that business is booming at the world's busiest spaceport.

Kennedy's expertise in constructing and operating the world's premier spaceport forms the foundation for a new Spaceport Technology Center (STC). The STC will encompass all new technologies supporting new and proposed spaceports on Earth and on other planets.

One Kennedy team is developing the concepts and the architecture of future spaceports. A major element of the work is to optimize ground operations for launch vehicle processing, to help lower the overall costs of space transportation. An interactive design tool is under development, to help spaceport planners assess interactions between needed ground infrastructure and flight systems. This activity, and other work, is being pursued to capture global launch capabilities and help focus future Kennedy roles in meeting NASA and customer requirements. With results in hand, Kennedy officials expect to foster new initiatives, upgrade existing systems, and respond to the technology needs of customers using the spaceport. Given movement toward new generations of launch vehicles, including reusable space transportation, taking a long look into the future is prudent.

Critical to future spaceports is environmental protection. The center occupies 140,000 acres of land and water on Merritt Island. Just a small fraction of Kennedy's land area is utilized for space operations. The remaining acreage is a wildlife refuge and national seashore. Meeting space launch demands, while being effective stewards of the precious natural environment, is an important challenge at the center, one that is being met by effective monitoring and protection of the area's environmental quality. Kennedy is committed to environmental leadership, and is making important strides in ecosystem management and modeling, environmental remediation, and remote sensing applications.

Kennedy is playing a significant role in the scientific advancement of bioregenerative life support systems for long duration spaceflight. Center researchers are engaged in ground-breaking studies in utilizing biological systems, life support systems, and microbial ecology. New technologies in lighting, nutrient delivery, microbial monitoring, closed chamber construction and control, gas exchange, and crop productivity are leading to future designs for life support systems. Production of edible crop biomass and the processing of other biomass and waste are enabling technologies for future human space travelers.

Kennedy is also supporting NASA's long-term evaluation of the effects of microgravity on plant microbial systems. Kennedy's work in this area of biological systems has already led to important partnerships with academia and industry in advancing information in biological systems and their management and commercial product development.

Kennedy Space Center workers rotate the International Space Station's Node 1 and Pressurized Mating Adapter-1.

Handling complex space launchers and satellites demands sharp skills and the ability to integrate and test. This operational knowledge and expertise at Kennedy is helping to master the creation of "what if" software, that is, highly intelligent computer test software that reduces the time of engineers to configure, operate, and interpret the results of tests. Advanced simulation techniques, including virtual reality, are evolving to ensure the role of mission processing is as cutting-edge as feasible.

Also supporting Kennedy's collective vision to foster the birth of the Spaceport Technology Center is the new cryogenics testbed, which is a partnership with industry and academia. Formed under a reimbursable Space Act Agreement, Kennedy's extensive knowledge regarding cryogenics can be harnessed for both space and down-to-earth initiatives.

Kennedy has long used cryogenic liquids as launch vehicle propellants. In the private sector, there are many applications for such super-cold liquids. Biology and medicine use liquid nitrogen for preservation and storage of human and animal cells and tissues, as well as for the destruction of cancer tissue. Hospitals use superconductive magnets cooled with liquid helium for magnetic resonance imaging (MRI). Also, the food industry uses liquid nitrogen for freezing and long-term storage. Kennedy's know-how in cryogenics has also spurred the development of new thermal insulation materials and methods for better overall performance of cryogenic containers and piping systems. The center's cryogenics testbed is viewed as an important step to promote international excellence in cryogenic testing, training, and education.

Roy D. Bridges, Jr., Kennedy director, foresees a productive future for America's premier spaceport. "Kennedy Space Center is in a unique position to carry the U.S. space program into the next century. Our work force has no equal in the launch and payload processing business, and our ability to look to the future and prepare for the natural evolution of missions and technology will ensure that we are strategically positioned to respond to those needs," Bridges says.

What that future might entail is thrilling, he adds. "While we look forward to preparing for the on-orbit assembly and operation of the International Space Station, we also eagerly embrace the challenge of what lies beyond, whether it be a return human mission to the Moon or the first crewed expedition to Mars. We'll be ready." ❖

The Cassini spacecraft is lowered onto its launch vehicle adapter in Kennedy Space Center's payload Hazardous Servicing Facility.

STS-80 lands at Kennedy Space Center's Shuttle Landing Facility.

Langley Research Center
NASA Headquarters and Centers

Langley Research Center is NASA's Center of Excellence in airframe systems and leads in airborne systems, structures and materials, aerodynamics, and mission and systems analysis. Located in Hampton, Virginia, Langley's unique research facilities include over 40 wind tunnels.

A community of Langley researchers is honing the technologies that enable aircraft to fly faster, farther, safer, and to be more maneuverable, quieter, less expensive to manufacture, and more energy efficient.

A new aviation safety initiative supported by Langley is the development of Aviation Weather Information (AWIN) systems. Airliners and smaller airplanes are one step closer to having up-to-the-minute, graphical weather displays in their cockpits. Research agreements between NASA and eight industry teams have been signed to bring about the AWIN initiative.

What is envisioned in the AWIN effort is a futuristic system that allows aircraft to be both a source and user of weather information. Airborne sensors would provide data for weather systems on the plane, on the ground, and in other aircraft. As part of the cockpit instrumentation, easy-to-read, real-time displays would show weather across the country, not just a limited number of miles ahead. Satellite and ground transceivers would move weather data to and from aircraft. Industry teams working on AWIN also propose other information tools, such as alarm systems or displays of suggested routes to help pilots better avoid potentially hazardous weather situations.

New designs are taking shape for a small aircraft transportation system to make personal air travel a safe, affordable transportation alternative.

AWIN is part of an overall NASA move toward improvements in aviation safety. Because of advances in the last four decades, commercial airliners are already the safest of all major modes of transportation. But with an accident rate that has remained relatively constant in the last decade and air traffic expected to triple over the next two decades, the U.S. government wants to prevent a projected rise in the number of aircraft accidents.

The long-running Advanced General Aviation Transport Experiment (AGATE) consortium continues to make impressive progress on a broad front. This partnership between government and industry is revitalizing the U.S. light airplane industry, strengthened by the NASA General Aviation Propulsion (GAP) program.

Langley, working with the Federal Aviation Administration (FAA), and U.S. industry, scored a range of recent accomplishments by way of the AGATE consortium. Among the achievements: AGATE safety advances in energy-absorbing and improved safety harness systems that improve crash protection; an AGATE-developed process that promises to speed materials certification by the FAA for a new aircraft design from two years to six months, promising up to $1 million in savings per plane; and reducing the cost of lightning protection for small airplanes from the current $5,000 per airplane to a projected goal of $500 or less by using airplane surface treatments.

Langley technology partnerships have been formed among companies, universities, and local government. Government-sponsored research at the center has moved into several commercially available products and techniques, including: a non-invasive method to measure, via ultrasonics, the signs of abnormally elevated intracranial pressure common in patients with head trauma; a field inspection technique to perform near real-time analysis of non-metallic inclusions in representative stainless steel samples; and ultrasonic sensor technology for accurately measuring the strain in bolts and fasteners.

Image illustrating vehicle separation from Pegasus booster during flight research.

In another example of a commercial partnership, Langley's Research and Technology Group has licensed technology for improving the performance, stability, and control of helicopters. Boundary layer research done at Langley, to study the flow of air around aircraft, has spawned a number of ideas. One aerodynamic device is called "tailboom strakes." This helicopter strake technology, developed by a team of NASA and Army researchers, counteracts a single-rotor helicopter's natural tendency to turn due to torque. The result is improved pilot control of the helicopter. NASA has licensed the technology for commercial sale to Boundary Layer Research, Inc., Everett, Washington (see p. 51).

Moving beyond technologies to improve helicopter stability all the way to hypersonic velocities is part of Langley's research agenda. Hyper-X hypersonic propulsion hardware is to undergo high-speed ground tests at Langley, which is also managing the program. Revolutionary Hyper-X vehicles are to carry "air-breathing" engines, burning oxygen that is scooped from the atmosphere. Hyper-X program managers hope to fly vehicles at speeds 10 times the speed of sound to altitudes of 100,000 feet.

Studies led by Langley concerning global warming may lead to more accurate climate predictions. PICASSO-CENA is a collaboration between Langley, France, and industry. PICASSO-CENA, also known as the Pathfinder Instruments for Cloud and Aerosol Spaceborne Observations-Climatologie Etendue des Nuages et des Aerosols, will profile the vertical distribution of clouds and aerosols. While these measurements are being taken, another device will simultaneously image the heat emission of the atmosphere.

Data from PICASSO-CENA, to be lofted in 2003, along with NASA's Earth Observing System satellites, will help piece together 3-D images of the atmosphere. The role of clouds and aerosols in Earth's climate can be better understood through such spaceborne instruments. These types of measurements should provide the scientific basis for understanding the dynamics and energetics of Earth's atmosphere on a short-term weather basis, perhaps leading to long-term climate forecasts. ❖

Color graphic model of the Hyper-X.

Marshall Space Flight Center
NASA Headquarters and Centers

Engineers at Marshall's Huntsville Operations Support Center (HOSC) perform analysis of activities during a Microgravity Science Lab simulation.

As NASA's lead center for space transportation systems development, Marshall's talents are well suited to pursue a mix of next-generation space transportation.

Marshall Space Flight Center is NASA's premier organization for developing space transportation, advancing propulsion concepts, as well as carrying out cutting-edge microgravity and optics research. Situated on 1,800 acres in Huntsville, Alabama, Marshall has played a leadership role in such efforts as developing the Saturn rockets used in the Apollo lunar landing program and managing the construction of the Skylab space station, the Space Shuttle Main Engine, as well as the powerful Hubble Space Telescope.

As NASA's lead center for space transportation systems development, Marshall's talents are well suited to pursue a mix of next-generation space transportation. "We are going on to revolutionary new propulsion technology and have an opportunity to perhaps, in some ways, relive those old days of being pioneers again," explains Arthur G. Stephenson, director of Marshall.

Marshall is managing the X-33, X-34, and Future-X programs. This trio of programs is tackling the difficult task of lowering the cost of low-Earth orbit, first from $10,000 to $1,000 per aircraft pound, then to as low as $100 per aircraft pound. As the flagship technology demonstrator for reusable launch vehicles (RLVs), under construction by the Lockheed Martin Skunk Works, the X-33 is to help cut the expense of access to space. This unpiloted, wedge-shaped rocket is slated to begin suborbital flight tests in mid-2000. It will reach an altitude of up to 50 miles and accelerate to between 13 to 15 times the speed of sound. As many as 15 flights of the X-33 are planned.

Another reusable, suborbital vehicle is the X-34. Built for air-launch and able to rocket as high as 50 miles in altitude, the X-34 is to reach speeds of up to eight times the speed of sound. This vehicle will demonstrate the ability to fly through inclement weather, land horizontally at a designated landing site, and safely abort a flight. The X-34 is to be capable of performing 27 flight tests within a period of one year, at an average recurring cost of approximately $500,000, and of demonstrating operations with a small work force. It will also demonstrate a 24-hour turnaround between two flights. The X-34 is to begin a series of test hops in late 1999. Powering the X-34 is the Fastrac engine, designed and developed by Marshall engineers. Orbital Sciences Corporation of Dulles, Virginia, is developing the X-34 under Marshall contract.

The Marshall-managed Future-X initiative encompasses the X-37, an unpiloted orbital craft that would autonomously land after re-entry. Marshall is working with the Boeing Phantom Works at Seal Beach, California, on the X-37 vehicle. This program is a key step toward achieving NASA's goal of a $1,000 per pound cost, or lower, to place spacecraft into orbit. A follow-up concept is the 21st century Spaceliner 100, a

Marshall-designed vehicle that could feature revolutionary air-breathing rocket engines, which would be catapulted skyward from a magnetically levitated launch rail system. Also under study at Marshall are beamed energy, nuclear, and solar electric propulsion ideas, as are pulse detonation engines, and high-energy propellants.

Marshall is responsible for the Michoud Assembly Facility in New Orleans, Louisiana. At this site, the Space Shuttle's huge external tank is manufactured. The center also is responsible for the assembly and refurbishment of the Shuttle's solid rocket motors, which is completed at the Kennedy Space Center in Florida.

Microgravity research is also underway at Marshall, with scientists partnered with the scientific community and commercial industry to study the effects of gravity on biological, chemical, and physical systems. NASA uses ground-based research and experiments on the Space Shuttle and the upcoming International Space Station to explore the impact of microgravity on processes. That research is offering insight into improving commercial products, including crystals, metals, ceramics, glasses, and biological materials. Marshall scientists believe that the opportunities offered by microgravity science are vast.

Engineers at Marshall specialize in optical systems for NASA's orbiting telescopes, such as the Chandra x-ray observatory and the future Solar X-ray Imager, developed to provide continuous, near real-time images of the Sun. Marshall expertise is also guiding optical work on NASA's Next-Generation Space Telescope (NGST). Set for launch around 2007, NGST will study the birth of the first galaxies, the shape and fate of the universe, the formation of stars and planets, the chemical evolution of the universe, and the nature of dark matter. Beyond NGST, ultra-lightweight and deployable mirrors for advanced space telescopes are being investigated.

Marshall is making significant contributions to the International Space Station program. Making use of manufacturing facilities provided by Marshall, Boeing has supplied major station elements, like the Unity connecting node and the Destiny laboratory module. Marshall is home to the Payload Operation Integration Center, the science operations gateway to the Space Station. Thanks to a Telescience Resource Kit developed at Marshall, scientists on the ground can remotely monitor and operate their Space Station experiments. Yet another task undertaken at Marshall is the water-recycling and oxygen-generating systems for the U.S. segment of the Space Station.

From air-breathing rocket engines to beamed energy, space tether research, and microgravity science on the International Space Station, Marshall is maximizing its scientific and technological talent in preparation for the next millennium. ❖

An engineer at one of Marshall's vacuum chambers tests a microthruster model. Data gathered from the testing provide information on propulsion in microgravity.

Research to lower the cost of thrust chamber assembly is conducted by a Marshall scientist to significantly reduce the costs associated with thrust chamber/injector development and fabrication.

Stennis Space Center
NASA Headquarters and Centers

The rumble and roar of rocket engine testing has long been a mainstay activity of the John C. Stennis Space Center. Arrays of Stennis propulsion test facilities are in constant use, and for good reason. Mississippi-based Stennis is NASA's lead center for testing large propulsion systems. Indeed, every Space Shuttle astronaut rides on rocket engines tested at Stennis. In October 1998, center engineers conducted the 2,000th test of a Space Shuttle Main Engine.

Building on its role in engine and vehicle testing, which spans 30 years and dates back to the Apollo lunar landing program, Stennis is now helping to shape the future by assembling and testing the RS-68 engine and first stage common booster core for Boeing's new Delta 4 rocket. The RS-68 is the world's largest liquid hydrogen and liquid oxygen engine. By way of a NASA and industry partnership, propulsion test facilities at Stennis are tapped for commercial use. Delta 4 is an Evolved Expendable Launch Vehicle (EELV), a U.S. Air Force program that will produce the next generation of unmanned launch vehicles for both civilian and military use.

Another Stennis rocket engine test stand has been converted and modified to test the X-33's linear aerospike engines, the propulsion end of a partnership between NASA and industry to develop Reusable Launch Vehicles (RLVs), designed to dramatically cut the cost of putting payloads into Earth orbit. A Stennis team began tests on critical elements of the aerospike motors, leading up to full-scale engine tests in 1999. The X-33 is a half-scale forerunner of a commercial version of an RLV, called VentureStar, which is to be built by the Lockheed Martin Skunk Works of Palmdale, California. VentureStar development will take place early next century.

Other major engine development projects are also underway at Stennis. Work is underway to ready for flight the 60,000-pound thrust engine, called Fastrac. A single-stage main engine, Fastrac burns a mixture of liquid oxygen and kerosene, powering the X-34 and the first stage of a small rocket booster for NASA's Low-Cost Technologies program. Also, a 250,000-pound thrust hybrid rocket motor is under evaluation. Hybrid rocket motors use an environmentally safe, rubberized fuel and a liquid oxidizer. In this hybrid test project, NASA and Lockheed Martin are conducting work through a Space Act Agreement, a partnership where both parties bring resources to the arrangement and both benefit from the effort.

A Space Shuttle Main Engine test lights up the night sky at Stennis Space Center.

Stennis is also working with small businesses under the Small Business Technology Transfer Program, focusing on a high-accuracy modular thrust measurement system, which is utilizing magnetic bearings, along with looking into hydrocarbon-fueled rocket engine health monitoring by laser-induced breakdown spectroscopy.

While Stennis is busily working on rocket engines that hurl payloads less expensively up above Earth, the center is also engaged in commercial remote sensing activities that look downward from space to scan the planet.

Stennis is NASA's lead center for commercial remote sensing within the space agency's Earth

The Commercial Remote Sensing Program at Stennis assists numerous companies to enhance their competitiveness.

Science Enterprise. The emerging remote sensing industry is fast becoming a potential multibillion-dollar force in the U.S. economy. In this regard, Stennis' Commercial Remote Sensing Program (CRSP) office has partnered with industry to help develop remote sensing technology. Remote sensing is the ability to acquire and produce images of specific areas of Earth using sensors mounted on aircraft or satellites. Benefits include: determining the best time to irrigate and fertilize crops; finding the most appropriate routes for highways; better planning in the placement of utility lines; and assessing environmental damage caused by oil spills and natural disasters.

One project underway at CRSP, which began in 1998, is a phased $50 million Science Data Buy, where 15 terabytes of data are being purchased from 5 companies. This data will be available for global environmental research within NASA's Earth Science Enterprise. In the future, NASA will continue to purchase data from the commercial sector and work to understand scientific needs and industry concerns. By purchasing data from the private sector, instead of developing, building, and launching new NASA satellites, the space agency may be able to conduct and expand its scientific investigations at a much lower cost, while encouraging the growth of this economic area. CRSP is also providing the remote sensing community with a comprehensive array of artificial and natural ground targets, to help test commercial airborne and spaceborne remote sensing systems against performance specifications and customer needs.

In 1998, Stennis' Earth System Science Office continued a study of Louisiana's Barataria Bay, researching the role of the bay in the global carbon cycle. This research is expected to yield new data with regard to shrimp and oyster production in the bay. Furthermore, this study complements remote sensing studies on the health of Lake Pontchartrain and the occurrence of red tide along the Louisiana coast.

Stennis' Earth Observations Commercial Applications Program works with business partners to demonstrate the market effectiveness of new remote sensing products. This is exemplified by recent projects to evaluate the commercial potential of synthetic aperture radar, as well as uses and benefits of hyperspectral data.

"As we face the changing times ahead and approach the new millennium, we intend to build on the accomplishments of a great foundation with a renewed focus," explains Stennis director Roy S. Estess. "Stennis Space Center will continue partnering with industry, government, and academia to provide our nation a return on its investment in the human exploration and development of space, as well as in our national defense, economic competitiveness, and study of the environment." ❖

Jet Propulsion Laboratory
NASA Headquarters and Centers

The Goldstone Deep Space Communications Complex, located in California's Mojave Desert, is one of three complexes in NASA's Deep Space Network (DSN).

Marvelous images from such far away locales as Mars, Jupiter, and Saturn have been made possible by the engineering and scientific minds at the Jet Propulsion Laboratory (JPL). Just as impressive are the missions JPL is flying or planning to take NASA into, what JPL director Dr. Edward Stone refers to as the "third era of exploration"—a time of sending probes into deep space often for detailed exploration or to return samples to Earth.

One of NASA's leaders in the agency's Space Science Enterprise, JPL is managed by the California Institute of Technology. JPL is NASA's Center of Excellence in deep space systems. This center also manages the worldwide Deep Space Network of radio dishes that are placed in California, Spain, and Australia.

JPL engineers are blueprinting missions for the robotic exploration of Mars, Jupiter's moon Europa, distant Pluto, the Sun, and numbers of comets and asteroids. A step-by-step build-up of technological capability has also started, with the aim of detecting and imaging Earth-like planets that circle stars light-years away. JPL also develops and flies instruments and satellites that observe the environment of Earth. Additionally, the laboratory uses its technological expertise for many other customers and partners both within and outside of NASA.

A good example of a recent cutting-edge JPL mission is that of the Deep Space 1. Launched into space in October 1998, Deep Space 1 took a host of advanced technologies around the test track.

Armed with advanced solar arrays, and carrying many other new technologies, including several in communications, microelectronics, and spacecraft structures, Deep Space 1 also validated the first-ever use of an ion propulsion system for primary propulsion in deep space. Fueled by xenon, the ion engine produces a small, but constant thrust for hours on end while being 10 times more fuel efficient than chemical onboard propulsion systems. From late November through the end of the year, the ion engine chalked up nearly 720 hours of engine thrusting and was still going strong. Deep Space 1's validation of its ion engine is viewed by many to be one of NASA's biggest breakthroughs ever.

JPL teams continue to improve on such technologies as the Quantum Well Infrared Photodetector, one of the world's most highly sensitive infrared cameras at long wavelengths. Work on the Active Pixel Sensor is being furthered, enabling video cameras to be reduced to the size of a chip coupled with optics, while using only one-hundredth the power of standard CCD cameras.

Sponsored by JPL's Technology Applications Programs (TAP), important milestones are being met in projects pursuing "Global Positioning System on a chip" technology, Millimeter Integrated Circuit low-noise amplifiers, and a submillimeter sensor to measure ice. TAP-developed technologies range from sensors to support far-infrared missions, to electronic components for detection needs in the microwave and submillimeter wave spectral regions. These technologies, and others developed at JPL, will enable such missions as the First Infrared and Submillimeter Space Telescope (FIRST), and contribute to the European Space Agency's Rosetta mission to a comet.

An artist's rendition of the Mars Polar Lander.

TAP is also developing Lithographie Galvanoformung Abformung (LIGA) grids for NASA's High Energy Solar Spectroscopic Imager (HESSI) mission to be launched in 2000. HESSI will provide full Sun spectral images to help discern a number of solar secrets. LIGA/thick film lithography is a technology with an impressive future, making possible specialized arrays for miniature mass spectrometers, miniature ion traps for mass spectroscopy, tunable miniature inductors and capacitors for power and communications applications, as well as micro-sized propulsion, power sources, and pumps. In the case of HESSI, the LIGA microfabrication technique reduced both spacecraft size and the mission's cost.

As part of its current proof-of-concept phase, the Viewing Imager Gimbaled Instrumentation Lab & Analog Neural Three-dimensional processing Experiment (VIGILANTE), has moved to completion phase. VIGILANTE is a machine vision instrument that combines several sensors in order to recognize specific targets in real-time, without the aid of the human eye. The key is a new, JPL-developed, sugar cube-sized processor built on neural networking principles.

JPL's Space Inflatable Technology Program is moving toward center stage. This innovative program is bringing closer the day when huge radar and communication dishes circle the Earth, enormous solar sails slip through the vacuum of space, and giant sunshades cool down high-tech infrared sensors that peer deep into the universe. Accomplishment in this area included the successful testing of a half-scale model of the inflatable sunshield for the Next-Generation Space Telescope (NGST). Also constructed were 42-foot-long inflatable thermoset composite booms for future applications on solar sails. Inflatable solar arrays are now part of the JPL's Deep Space 4 mission to land atop a comet.

Be it ultra-miniature instruments or huge inflatable structures, JPL scientists and engineers have shown technology advancements come in all shapes and sizes. ❖

An artist's rendering of NASA's Quick Scatterometer (QuikScat) that measures winds over the ocean surface.

Technology Transfer and Commercialization

In 1958, Congress mandated the National Aeronautics and Space Administration (NASA) to disseminate the greatest amount of information possible resulting from its research and development efforts to the public.

Forty-one years later, NASA's Commercial Technology Network, along with each NASA field center's Commercial Technology Office, is extending the reach of NASA technology into the everyday lives of the American public. The wide network of organizations focuses on transferring technology and provides a vast array of products and services geared to enhance and further the global competitiveness of U.S. industry.

The Commercial Technology Division at NASA Headquarters and the Commercial Technology Offices at each field center serve as gateways to accessing the cutting-edge research and technology available for transfer and commercial use. Each year, NASA *Spinoff* highlights this technology transfer.

John H. Glenn Research Center at Lewis Field

Technology Transfer and Commercialization

NASA's Glenn Research Center is located in Cleveland, Ohio.

Nineteen hundred ninety-nine has been a year of esteemed recognition for NASA's John H. Glenn Research Center at Lewis Field. Three individuals associated with the research facility were honored for their contributions in advancing the benefits of research in aeronautics and aerospace. George W. Lewis and John H. Glenn were recognized for their outstanding achievements in the May 7, 1999, renaming ceremony of the research center, formerly called the Lewis Research Center. George Lewis, who died in 1948, was represented by his grandson at the ceremony. On April 21, 1999, the Federal Laboratory Consortium (FLC) named Glenn Research Center director Donald J. Campbell the Laboratory Director of the Year (1998) for Technology Transfer.

Director Campbell, Glenn's first African-American center director, was selected to receive the award in recognition of his successful efforts to broaden the commercialization of Glenn's technologies. Under Campbell's leadership, a number of distinguished technology transfer initiatives have flourished. Among these initiatives are Glenn's Commercial Technology Office (CTO) and the Lewis Incubator For Technology (LIFT), both established to help entrepreneurs and start-up companies gain financial and marketing assistance in commercializing NASA-developed technologies.

The Glenn Commercial Technology Office

Glenn has taken part in a number of research and development projects that have advanced technologies in our everyday lives. To further these advancements, Glenn created its CTO. The field center's CTO was introduced in 1995 as Glenn's main liaison to non-aerospace companies and external organizations involved in economic development, technology transfer, and commercialization. Glenn's technology transfer program contributes significantly to the ability of Northeast Ohio companies to compete in global markets.

Technology transfer provides area companies with innovative technologies and facilitates the solutions to real-world, commercial problems by Glenn scientists and engineers. The resulting benefit is the advancement or creation of products, processes, and services.

An example of Glenn's CTO efforts in technology transfer is the Embedded Web/Tempest Training Workshop and Developer's Kit, which offers a complete software system that can be adapted to let users control and monitor, via the Internet, devices with embedded microprocessors. Using various programming languages, the technology allows scientists to remotely control and monitor experiments on space missions, including those on the International Space Station. A wide variety of commercial applications for this technology are being realized. Companies can control equipment and machinery in remote manufacturing plants by connecting their machines to the Internet and feeding data directly into databases and analysis programs. Landlords can monitor the heat and use of office buildings from miles away. Stores and museums can control cameras to view special merchandise or displays. Among the benefits of this technology is the ability to improve products and services and to integrate production equipment with other corporate systems. The Embedded Web/Tempest project recently received the 1999 R&D 100 Award and was the 1998 NASA Software of the Year winner.

NASA has been involved in the development and commercial transfer of many of the revolutionary and evolutionary technologies that define this century. "Looking at the same time frame, it is unlikely that technology would have reached its current level of maturity without NASA," says Campbell. Glenn has helped hundreds of companies

Director Donald J. Campbell, head of the Glenn Research Center, has been recognized for his outstanding efforts in transferring NASA technology.

turn high-tech research into marketable products, and Glenn's CTO serves as the hub for making these achievements possible.

The Glenn Research Center's LIFT Program

Glenn's LIFT program is a business incubator designed to nurture new and emerging businesses with the potential to incorporate technology developed by Glenn. LIFT offers high quality laboratory and office space at attractive rates, along with a wide variety of business and technical support services to increase the success of participating companies. Access to the outstanding technology and support resources of Glenn through actively developed links is one of the strongest benefits of the program.

Funded by NASA and the Ohio Department of Development, LIFT is a cooperative effort among Glenn, the Ohio Department of Development, Enterprise Development, Inc., the Great Lakes Industrial Technology Center (GLITeC), and BP America. Its primary objectives are to create new products, services, companies, and jobs in Ohio and to increase the commercial value of technology developed at Glenn.

LIFT's executive director, Wayne Zeman, explains, "The whole idea behind LIFT is to make it easier for the entrepreneur. LIFT is like a cocoon of technical and business support around start-ups." Glenn has documented thousands of technologies in areas such as high-performance materials, advanced communications systems, and cutting-edge electronic sensors available for use in "non-NASA" industries.

Glenn Research Center Partnerships for Technology Transfer

Glenn's newly created Garrett Morgan Commercialization Initiative is another example of the field center's efforts to establish effective technology transfer partnerships. The Garrett Morgan Commercialization Initiative, named after the African-American inventor and Cleveland resident responsible for the traffic signal and the safety helmet/gas mask, helps to increase the competitiveness of small and disadvantaged businesses in Ohio and the Great Lakes region through the use of NASA technologies. The program offers companies marketing assistance, technical expertise, and help in planning Small Business Innovation Research (SBIR) proposals, among other services.

Glenn's partnership with NASA's Midwest Regional Technology Transfer Center and GLITeC plays a significant role in the successful transfer of NASA technology. GLITeC is charged with improving the competitiveness of industry through the efficient application of technology and related capabilities from Glenn and other federal laboratories.

An example of Glenn's innovative efforts with GLITeC is the Glennan Microsystems Initiative, named after a former president of Case Western Reserve University and the first NASA administrator. The Glennan Microsystems Initiative is a NASA-financed project to advance the revolutionary new technology known as microelectromechanical systems (MEMS), the development of which involves miniaturized electrical and mechanical

John H. Glenn Research Center at Lewis Field *Continued*

Technology Transfer and Commercialization

devices, such as sensors, motors, gears, valves, and microprocessors that work together to perform a task. The devices could be used to improve high-speed optical switches, create more efficient fuel atomizers, control air pressure in "smart tires," or even create better barcode scanners.

Glenn Technology Transfer Successes

Glenn's technology transfer successes are continually contributing to the improvement of all aspects of society. From cleaner and more efficient transit buses to art restoration and improvements in the understanding of diabetes, Glenn is dedicated to furthering the successful transfer, use, and commercialization of NASA technology to benefit the general public.

Glenn has managed a project to develop a transit bus, known as the Hybrid Electric Transit Bus, which uses a natural gas turbine to produce electricity. The electricity generated, in turn, powers a variable-speed drive train. Lincoln Electric, Teledyne Ryan Aeronautical, Bowling Green State University, and the Cleveland Regional Transit Authority (RTA) have also participated in the project. This hybrid bus could double the fuel economy of inner city buses, while reducing exhaust emissions, lowering noise, and reducing maintenance costs. The state-of-the-art bus promises emissions measuring one-tenth of Environmental Protection Agency standards. Unlike conventional buses that waste fuel while idling in traffic, the Hybrid Electric Transit Bus continues to run at near peak efficiency while stopped. The power is stored and used when the bus needs a boost, such as when climbing steep hills.

Glenn's technological expertise has also gone into helping the sailing team *America True* gain a significant technological advantage over the competition. *America True* formed an alliance with Glenn for its race for the America's Cup, sailing's oldest trophy. The race will begin in February 2000 in Auckland, New Zealand. Engineers at Glenn supported the *America True* design team. An official Space Act Agreement summarized the partnership, calling for research consultation on sail and mast design. Using high-fidelity computational analysis tools, currently in use by Glenn for propulsion applications, designers analytically tested the performance qualities of different sail and mast designs to select the optimal shapes and material configurations for the racing boat. The similarity between Glenn's jet engine performance simulations and the capabilities required for advanced boat engineers enables *America True* to benefit from NASA's experiences.

Glenn's technological developments have also extended into the creation of better televisions. A high-temperature, high-voltage, semiconductor called Silicon Carbide (SiC), developed by Glenn and delivering three times the power of conventional silicon devices, is helping to accelerate the use of high-definition television (HDTV). It also promises to bring cinema-quality pictures and compact disc sound to the United States and abroad during the 21st century.

Another transferred NASA technology is one that was originally developed for space propulsion and simulation of the space environment. The technology, known as atomic oxygen, has made it possible to alter the surface of many materials through texturing, including artistic paintings. The process involves texturing or removal of organic layers with a low-energy beam of oxygen atoms. If the organic layer is on paint or another surface that is less reactive, the organic layer can be removed, without harming the underlying paint. Glenn has aided the Cleveland Museum of Art in some of its preservation and restoration efforts through this unique atomic oxygen technique.

Under the direction of Glenn research, the Hybrid Electric Transit Bus promises cleaner emissions and greater efficiency in power use.

38   Center Spotlight

Glenn worked with the Cleveland Museum of Art on another research effort involving the examination of an ancient Egyptian art collection. The researchers wanted to determine how ancient craftsman mixed materials to create Egyptian Blue, the world's first synthetic pigment. Small samples taken from experimental reproductions were viewed with a scanning electron microscope, which magnified the samples by many thousands.

Another Glenn technology can be attributed to helping in the research and understanding of diabetes. A researcher at the Cleveland Clinic Foundation (CCF) developed a sensor pad for measuring the friction and pressure forces under a person's foot when walking. Glenn's Interactive Data Display System (IDDS), which allows data to be shown in 2-D and 3-D images, was installed at the CCF to be used in conjunction with this sensor pad. Using NASA's IDDS technology along with the sensor pad, data are mapped onto a grid with the outline of the patient's foot mapped over the data, assisting in visualizing the risk factors related to diabetic foot ulcerations. Glenn's IDDS is assisting the CCF in understanding a problem that accounts for 20 percent of all hospital admissions for diabetic patients. The information the sensor pad and the IDDS provide is contributing to the improvement and understanding of the factors that lead to skin breakdown and ultimately a reduction of the quality of life for diabetic patients.

Glenn Education Initiatives

Glenn is involved in a number of educational partnerships. The objectives of these partnerships involve sparking the interests of elementary, middle, and high school students. Glenn's aim is to get students involved and interested in the disciplines of mathematics and science, in an effort to help encourage the continued success of the country's advancements in engineering, aeronautics, and aerospace.

Each year, Director Campbell has been instrumental in providing hands-on educational experience to African-American and Hispanic students through the Science, Engineering, Mathematics, and Aerospace Academy (SEMAA), a collaborative effort between Glenn and Cuyahoga Community College in Cleveland, Ohio. Because of the program's success, SEMAA has become a model for a national education system, replicated in Dayton, Ohio; Detroit, Michigan; St. Louis, Missouri; and Washington, D.C.

The Glenn Research Center at Lewis Field is continuing its efforts to extend the reach of NASA technology in both the private sector and education. Through establishing partnerships with commercial development organizations and educational institutions and programs, Glenn is taking great strides to make NASA technology available for the benefit of everyone. ❖

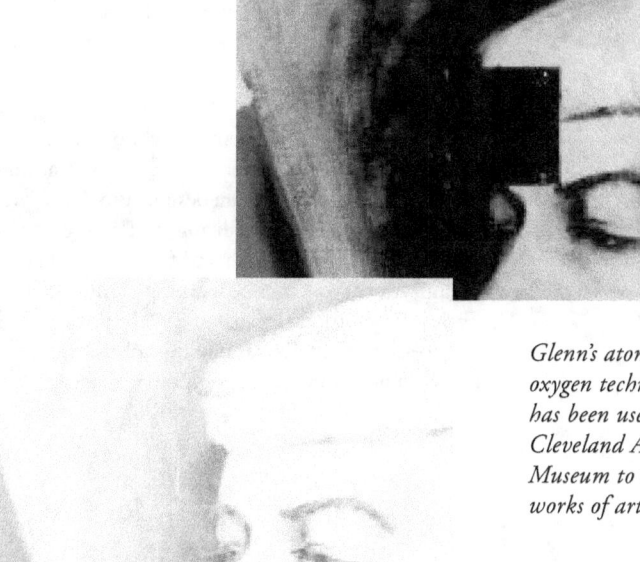

Glenn's atomic oxygen technology has been used by the Cleveland Art Museum to restore works of art.

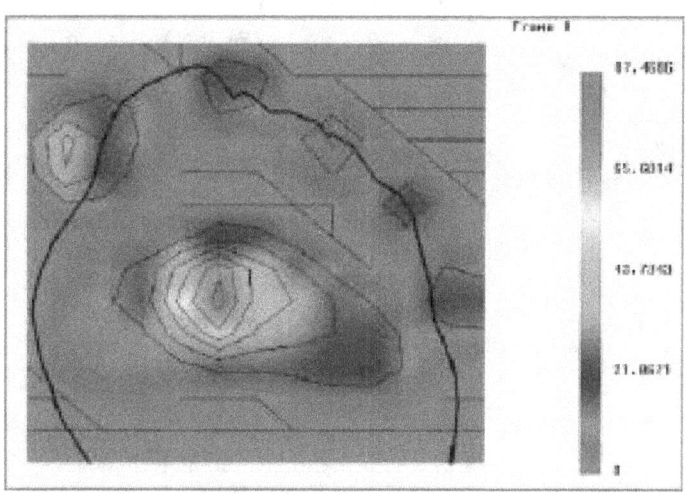

Glenn's Interactive Data Display System has been used in the study of diabetes-related problems.

NASA Technology Transfer and Commercialization

Technology Transfer and Commercialization

The NASA Commercial Technology Program sponsors a number of organizations around the country that are designed to assist U.S. businesses in accessing, utilizing, and commercializing NASA-funded research and technology. These organizations work closely with the Commercial Technology Offices, located at each of the 10 NASA field centers, providing a full range of technology transfer and commercialization services and assistance.

The National Technology Transfer Center is located on the campus of Wheeling Jesuit College in Wheeling, West Virginia.

Technology Transfer Network

A key player in the transfer of NASA technology is the **National Technology Transfer Center** (NTTC) <http://www.nttc.edu>. The NTTC, located at Wheeling Jesuit College in Wheeling, West Virginia, is an integrated resource for accessing federally-funded research and development and other information. The NTTC offers a free service, known as the Gateway, through which the private sector can maintain person-to-person contacts in the federal laboratory system. The Gateway provides a direct link to many resources, including technical reports, technologies, and facilities at NASA and other federal laboratories.

The NTTC is also responsible for administering **NASA TechTracS** <http://technology.nasa.gov>, which provides access to NASA's technology inventory and numerous examples of the successful transfer of NASA-sponsored technology for commercialization. TechFinder, the main feature of the Internet site, allows users to search the technologies and success stories, as well as submit requests for additional information. All NASA field centers submit information to the TechTracS database as a means of tracking technologies that have potential for commercial development.

Since their inception in January 1992, the six NASA-sponsored Regional Technology Transfer Centers (RTTCs) have helped U.S. businesses investigate and utilize NASA and other federally-funded technologies for commercial companies seeking new products, improvements to existing products, or solutions to technical problems. The RTTCs provide technical and business assistance to several thousand customers every year.

The network of RTTCs is divided up as follows:

Far West (AK, AZ, CA, HI, ID, NV, OR, WA):

The **Far West Regional Technology Transfer Center** (FWRTTC) <http://www.usc.edu/dept/engineering/TTC/NASA> is an engineering research center within the School of Engineering at the University of Southern California in Los Angeles. Using the Remote Information Service to generate information from hundreds of federal databases, FWRTTC's staff works closely with businesses and entrepreneurs to identify opportunities, expertise, and other necessary resources. The FWRTTC enhances the relationships between NASA and the private sector by offering many unique services, such as the NASA On-line Resource Workshop, NASA Tech Opps, and links to funding and conference updates.

Mid-Atlantic (DC, DE, MD, PA, VA, WV):

The **Mid-Atlantic Technology Applications Center** (MTAC) <http://oracle.mtac.pitt.edu/WWW/MTAC.html> is located at the University of Pittsburgh in Pennsylvania. MTAC has designed TechScout, a highly specialized set of matchmaking services, created to help companies locate technologies, as well as technical expertise, within NASA and the federal laboratory system. Close relationships with the Goddard Space Flight Center and the Langley Research Center allow MTAC to help the U.S. improve its competitiveness.

Mid-Continent (AR, CO, IA, KS, MO, MT, ND, NE, NM, OK, SD, TX, UT, WY):

The **Mid-Continent Technology Transfer Center** (MCTTC) <http://www.tedd.org/MCTTC>, under the direction of the Technology and Economic Development Division of the Texas

Engineering Service, is located in College Station, Texas. The MCTCC, which provides a link between private companies and federal laboratories, reports directly to the Johnson Space Center. The assistance focuses on high-tech and manufacturing companies that need to acquire and commercialize new technology.

Mid-West (IL, IN, MI, MN, OH, WI):

The **Great Lakes Industrial Technology Center** (GLITeC) <http://www.battelle.org/glitec>, managed by Battelle Memorial Institute, is located in Cleveland, Ohio. GLITeC works with industries primarily within its six-state region to acquire and use NASA technology and expertise, especially at the Glenn Research Center. Each year, over 500 companies work with GLITeC and its affiliates to identify new market and product opportunities. Technology-based problem solving, product planning and development, and technology commercialization assistance are among the services provided.

Northeast (CT, MA, ME, NH, NJ, NY, RI, VT):

The **Center for Technology Commercialization** (CTC) <http://www.ctc.org> is a non-profit organization, based in Westborough, Massachusetts. Covering New England, New York, and New Jersey, the CTC currently has seven satellite offices that form strong relationships with the Northeast industry. Operated by the CTC, the NASA Business Outreach Office stimulates business among regional contractors, NASA field centers, and NASA prime contractors.

Southeast (AL, FL, GA, KY, LA, MS, NC, SC, TN):

The **Southern Technology Applications Center** (STAC) <http://www.state.fl.us/stac> is headquartered at the University of Florida in Gainesville. Working closely with the Marshall Space Flight Center, the Kennedy Space Center, and the Stennis Space Center, STAC helps to spur economic development in each of the nine states in the southeast. To facilitate the transfer of NASA technologies and expertise, the three NASA centers and STAC formed the NASA Southeast Technology Alliance.

NASA Incubator Programs

Nine NASA incubators are included within this network of programs. They are designed to nurture new and emerging businesses with the potential to incorporate technology developed by NASA. They offer a wide variety of business and technical support services to increase the success of participating companies.

The **Ames Technology Commercialization Center** (ATCC) <http://ctoserver.arc.nasa.gov/atcc/atcc.html>, located in San Jose, California, provides opportunities for start-up companies to utilize NASA technologies. The center uses a lab-to-market approach that takes the technological output of Ames' labs and pairs that technology with appropriate markets to create and foster new industry and jobs. The incubator helps businesses and entrepreneurs find NASA technology with commercial potential, then provides access to a network of business experts in marketing, sales, high-tech management and operations, financing, and patent and corporate law. The ATCC also offers low-cost office space and other start-up services.

BizTech <http://europa.uah.edu/biztech/index.html> of Huntsville, Alabama, is a small business incubator, offering participating companies access to services at Marshall Space Flight Center laboratories for feasibility testing, prototype fabrication, and advice on technology usage and transfer. BizTech is sponsored by the Huntsville-Madison County Chamber of Commerce.

The **Emerging Technology Center** (ETC) <http://www.etcbaltimore.com/index2.html>, located in Baltimore, Maryland, is among the newest of the NASA-affiliated incubators. Partnering institutions include the Goddard Space Flight Center and area universities and colleges.

The **Florida/NASA Business Incubator Center** (FNBIC) <http://technology.ksc.nasa.gov/FNBIC> is a joint partnership of NASA's Kennedy Space Center, Brevard Community College, and the Technological Research and Development Authority. The mission of FNBIC is to increase the number of successful technology-based small businesses originating in, developing in, or that relocated to Brevard County. FNBIC offers support facilities and programs to train and nurture new entrepreneurs in the establishment and operation of developing ventures based on NASA technology.

The **Hampton Roads Technology Incubator** (HRTI) <http://www.hr-incubator.org> identifies and licenses NASA Langley Research Center technologies for commercial use. HRTI's mission is to increase the number of successful technology-based companies originating in, developing in, or relocating to the Hampton Roads area.

The **Lewis Incubator For Technology** (LIFT) <http://www.liftinc.org>, managed by Enterprise Development, Inc., provides outstanding resources

NASA Technology Transfer and Commercialization *Continued*

Technology Transfer and Commercialization

for technology and support to businesses in the Ohio region. Its primary objectives are to create businesses and jobs in Ohio and to increase the commercial value of NASA knowledge, technology, and expertise. LIFT offers a wide range of services and facilities to the entrepreneur to increase the probability of business success.

The **Mississippi Enterprise for Technology** is sponsored by NASA and the Mississippi University Consortium and Department of Economic and Community Development, as well as the private sector. The mission of the Enterprise is to help small businesses utilize the scientific knowledge and technical expertise at the Stennis Space Center. A significant part of this effort is Stennis' Commercial Remote Sensing Program (CRSP), which was formed to commercialize remote sensing, geographic information systems, and related imaging technologies.

The Hampton Roads Technology Transfer Incubator helps area companies in licensing NASA Langley Research Center technologies.

The **NASA Commercialization Center** (NCC) <http://www.nasaincubator.csupomona.edu/home.htm>, run by California State Polytechnic University, Pomona, is a business incubator dedicated to helping small businesses access and commercialize Jet Propulsion Laboratory (JPL) and Dryden Flight Research technologies.

The **UH-NASA Technology Commercialization Incubator** is a partnership between NASA's Johnson Space Center and the University of Houston. The incubator is designed to help local small and mid-sized businesses commercialize space technology. The University of Houston houses the program and provides the commercialization and research expertise of its business and engineering faculties.

Other organizations devoted to the transfer of NASA technology are the **Research Triangle Institute** (RTI) and the **MSU TechLink Center**.

RTI <http://www.rti.org>, located in Research Triangle Park, North Carolina, provides a range of technology management services to NASA. RTI performs technology assessments to determine applications and commercial potential of NASA technology, as well as market analysis, and commercialization and partnership development. RTI works closely with all of NASA's Commercial Technology Offices.

The **MSU TechLink Center** <http://www.montana.edu/techlink>, located at Montana State University-Bozeman, was established in 1997 to match the technology needs of client companies with resources throughout NASA and the federal laboratory system. TechLink focuses on a five-state region that includes Montana, Idaho, Wyoming, South Dakota, and North Dakota. Working closely with public, private, and university programs, TechLink provides ongoing support in the process of adapting, integrating, and commercializing NASA technology.

Affiliated Organizations, Services, and Products

To compliment the specialized centers and programs sponsored by the NASA Commercial Technology Program, affiliated organizations and services have been formed to strengthen NASA's commitment to U.S. businesses. Private and public sector enterprises build upon NASA's experience in technology transfer in order to assist the channeling of NASA technology into the commercial marketplace.

The NASA **Small Business Innovation Research** (SBIR) Program <http://sbir.gsfc.nasa.gov/SBIR.html> provides seed money to U.S. small businesses for developing innovative concepts that meet NASA mission requirements. Each year, NASA invites small businesses to offer proposals in response to technical topics listed in the annual SBIR Program Solicitation. The NASA field centers negotiate and award the contracts, and monitor the work.

NASA's SBIR Program is implemented in three phases:

• **Phase I** is the opportunity to establish the feasibility and technical merit of a proposed innovation. Selected competitively, NASA Phase I contracts last six months and must remain under specific monetary limits.

• **Phase II** is the major research and development effort, which continues the most promising of the Phase I projects based on scientific and technical merit, results of Phase I, expected value to NASA, company capability, and commercial potential. Phase II places greater emphasis on the commercial value of the innovation. The contracts are usually for a period of 24 months and again must not exceed specified monetary limits.

• **Phase III** is the process of completing the development of a product to make it commercially available. While the financial resources needed must be obtained from sources other than the funding set aside for the SBIR, NASA may fund Phase III activities for follow-on development or for production of an innovation for its own use.

The SBIR Management Office, located at the Goddard Space Flight Center, provides overall management and direction of the SBIR program.

The NASA **Small Business Technology Transfer** (STTR) Program **<http://sbir.gsfc.nasa.gov/SBIR.html>** awards contracts to small businesses for cooperative research and development with a research institution through a uniform, three-phase process. The goal of Congress in establishing the STTR Program was to transfer technology developed by universities and federal laboratories to the marketplace through the entrepreneurship of a small business.

Although modeled after the SBIR Program, STTR is a separate activity and is separately funded. The STTR Program differs from the SBIR Program in that the funding and technical scope is limited and participants must be teams of small businesses and research institutions that will conduct joint research.

The **Federal Laboratory Consortium** (FLC) for Technology Transfer **<http://www.fedlabs.org>** was organized in 1974 to promote and strengthen technology transfer nationwide. More than 600 major federal laboratories and centers, including NASA, are currently members. The mission of the FLC is twofold:

• To promote and facilitate the rapid movement of federal laboratory research results and technologies into the mainstream of the U.S. economy.

• To use a coordinated program that meets the technology transfer support needs of FLC member laboratories, agencies, and their potential partners in the transfer process.

The **National Robotics Engineering Consortium** (NREC) **<http://cronos.rec.ri.cmu.edu>** is a cooperative venture among NASA, the city of Pittsburgh, the state of Pennsylvania, and Carnegie Mellon's Robotics Institute. Its mission is to move NASA-funded robotics technology to industry. Industrial partners join the NREC with the goal of using technology to gain a greater market share, develop new niche markets, or create entirely new markets within their area of expertise.

The road to technology commercialization begins with the basic and applied research results from the work of scientists, engineers, and other technical and management personnel. The NASA **Scientific and Technical Information** (STI) **Program <http://www.sti.nasa.gov>** provides the widest appropriate dissemination of NASA's research results. The STI Program acquires, processes, archives, announces, and disseminates NASA's internal, as well as worldwide, STI.

The NASA STI Program offers users such things as Internet access to its database of over 3 million abstracts, on-line ordering of documents, and the NASA STI Help Desk for assistance in accessing STI resources and information. Free registration with the program is available through the NASA Center for AeroSpace Information (CASI).

For more than three decades, reporting to industry on any new, commercially significant technologies developed in the course of NASA research and development efforts has been accomplished through the publication of *NASA Tech Briefs* **<http://www.nasatech.com>**.

The monthly magazine features innovations from NASA, industry partners, and contractors that can be applied to develop new or improved products and solve engineering or manufacturing problems. Authored by the engineers or scientists who performed the original work, the briefs cover a variety of disciplines, including computer software, mechanics, and life sciences. Most briefs offer a free supplemental Technical Support Package (TSP), which explains the technology in greater detail and provides contact points for questions or licensing discussions.

Aerospace Technology Innovation **<http://nctn.hq.nasa.gov/innovation/index.html>** is published bimonthly by the NASA Office of Aeronautics and Space Transportation Technology. Regular features include current news and opportunities in technology transfer and commercialization, aerospace technology and development, and innovative research.

NASA Spinoff **<http://www.sti.nasa.gov/tto/spinoff.html>** is an annual print and on-line publication featuring current research and development efforts, the NASA Commercial Technology Program, and successful commercial and industrial applications of NASA. ❖

Commercial Benefits—Spinoffs

In order to assist NASA in its commercialization efforts of new, sophisticated, and versatile technologies, the agency's Technology Transfer Program was redefined and considerably broadened. The NASA Commercial Technology Network (NCTN) is a new mechanism combining many different services and organizations nationwide to promote commercialization.*

The NCTN has been essential in assisting commercialization of brand new technologies, enabling U.S. industry to remain globally competitive, while effecting benefits for all humankind at the same time.

The following pages demonstrate the applicability of technologies resulting from the many, often captivating, research and development activities at NASA, its transfer, and its commercialization by the engineering community. Wide-ranging outreach and partnership activities under the NCTN have again resulted in many new and diverse products and services—known as "spinoffs."

*For more information, see the NCTN website at <http://nctn.hq.nasa.gov>.

Blood Collection

Commercial Benefits—Spinoffs

Space travelers experience many physiological changes as they orbit Earth. An astronaut's body, once free of a 1-gravity pull, experiences a redistribution of body fluids. Proportionately more blood surges through the head, neck, and chest. Blood plasma volume is affected, and the number of red blood cells eventually decreases, leading to a form of space anemia. These effects and others are under study by NASA physicians to better appreciate how the human body reacts and adapts to microgravity and then readjusts to the Earth's gravity, once returned from space.

As part of these studies, NASA sought development of a device for the collection and real-time analysis of blood and other bodily fluids on missions without centrifugation. A method to collect and store such samples was invented by NASA and has been licensed to DBCD, Inc., of Webster, Texas. DBCD was formed in 1997 to commercialize the technology.

Under a patent licensing agreement from NASA, DBCD is now manufacturing a completely new range of blood separation products. These products incorporate the patented separation technology developed by NASA engineers. In May 1998, DBCD released its first product, the ProSeptor™, a blood drop device, and sales have begun. The ProSeptor™ 200, capable of collecting 150 microliters of serum, was also released.

The patented method and technology separate a relatively large volume of blood into cellular and acellular fractions without the need of a spinning centrifuge to accomplish this division. DBCD's ProSeptor™ products can provide serum or plasma from whole blood volumes of 20 microliters to 4 milliliters. These devices have a fibrous filter with a pore size of less than about 3 microns and are coated with a mixture of mannitol and plasma fraction protein. This coating causes the cellular fraction to be trapped by the small pores, leaving the cellular fraction intact on the fibrous filter. Meanwhile, the acellular fraction passes unaltered through the filter for collection from the serum sample collection chamber.

DBCD devices permit the collection of serum, anywhere, anytime, and from any species. No longer are heavy centrifuges required to be transported to remote sites, nor do police detectives have to hurry back to a lab to avoid sample spoilage. Expensive overnight parcels can be replaced by using the DBCD equipment.

Now manufacturing a range of blood separation products, DBCD services include customizing a blood separation system for a customer's instrumentation. DBCD's processing separator system equipment allows a user to collect, in only seconds to minutes, quantities of serum/plasma from whole blood that are of excellent quality.

ProSeptor™ 200 is designed to collect around 150 microliters of serum from 0.5 milliliters of blood in a laboratory or remote setting. The serum, which is collected on an easily removed collector, can be squeezed off for immediate analysis, frozen, or dried for later study. An Express Pouch is a specifically designed plastic pouch with storage material incorporated. The ProSeptor™ device or collection layer of the ProSeptor™ 200 can be placed in this pouch to dry and then be mailed to the laboratory for later analysis.

"DBCD has had great response from the community, diagnostic companies, and distributors regarding our products," says Eden Fields, president of DBCD. "NASA provides a great environment for development of products that can be beneficial to society. This technology is a good example of such a project," Fields adds. ❖

ProSeptor™ is a trademark of DBCD, Inc.

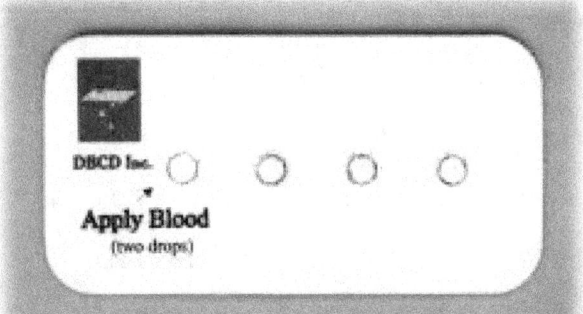

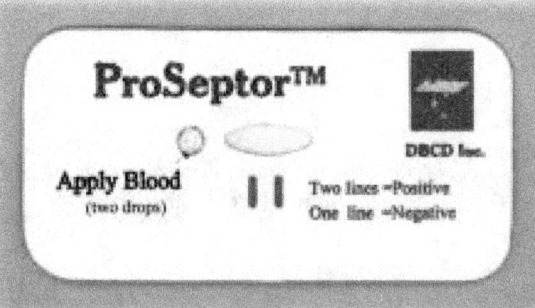

DBCD, Inc., has developed portable blood collection products that separate blood without the need of a spinning centrifuge.

The Eyes Have It
Commercial Benefits—Spinoffs

Serious buffeting that may occur when flying at high speed and low altitude can create visual blurring, hindering a pilot's ability to fly safely. NASA-funded research into the muscular physiology of eye vibration was the impetus for an optical instrument, now in medical use, that accurately measures eye movements.

Starting as far back as 1965, NASA's Ames Research Center contracted with the Stanford Research Institute (SRI) International of Menlo Park, California, to model the human visual-accommodation system, starting directly with the retinal image. This research was directed at how the major muscle systems control the human visual system.

Although a considerable amount of experimental information had been gathered regarding each of these systems, there was scant understanding of how control was actually achieved. One question that had gone unanswered was the voluntary and involuntary aspects of the human eye.

Several years of incremental NASA funding helped foster SRI-built eyetracking devices. These tools were able to anticipate, track, and monitor involuntary ocular movement horizontally, vertically, and with respect to depth of field. This development did not go unnoticed by leading academic and research institutions as an important instrument for understanding the visual system.

Since 1988, Fourward Optical Technologies, Inc., of San Marcos, Texas, has been manufacturing and marketing the Dual-Purkinje-Image (DPI) Eyetracker under an exclusive license. Purkinje images are four reflections produced from the front and rear surfaces of the cornea and lens. By observing the movement of the first and fourth Purkinje images with the DPI Eyetracker, the direction of gaze over a large, two-dimensional visual field can be determined with great accuracy. The instrument operates with infrared light, requires no attachments to the eye, and is not disruptive to normal vision.

Since its introduction by SRI, the Eyetracker has gone through several generations of development. The price of the early device was cut by more than half, making it suitable for clinical medical use. Applications of the Eyetracker are impressive. In cases of ocular bleeding, lasers can be used to stem the flow. The Eyetracker makes it possible to accurately target these problem areas, increasing the accuracy and the effectiveness of the treatment. Also, various brain disorders can now be diagnosed, as the Eyetracker can zero-out the eye's involuntary movements during diagnosis. Working in concert with lasers, the Eyetracker can assist non-invasive determination of a patient's circulatory health. By accounting for the eye's natural involuntary movements, the Eyetracker enables the operator to monitor blood flow by accurately targeting the retinal capillaries for Doppler blood flow studies.

The Eyetracker has a pointing accuracy on the order of one minute of arc and a response time on the order of one millisecond. By attaching Fourward Optical Technologies' Infrared Optometer to the device, continuous measurement of eye focus is possible, producing a 3-D Eyetracker. Eyetracker units are installed in over 11 countries around the globe, in addition to 30 or more sites in the United States.

Current and potential applications of the Fourward Optical Technologies' DPI Eyetracker include analysis of visual perception, mapping retinal features, neurological investigation, drug evaluation, and even the analysis of advertising material. No doubt the company has an eye toward the future. ❖

The Eyetracker, built by Fourward Optical Technologies, can precisely monitor eye movements, benefiting the medical field in various ways, from neurological diagnosis to blood flow monitoring.

Health and Medicine 47

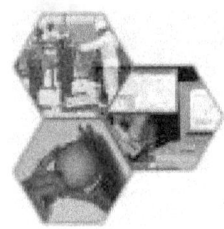

Just One Look

Commercial Benefits—Spinoffs

EMPAS is seen as a form of predictive artificial intelligence.

A line of advanced engine-monitoring systems, using the latest technology in graphic analog and digital displays can make a pilot's life in the cockpit much easier.

Working under a Small Business Innovation Research (SBIR) agreement with Langley Research Center, Vision Microsystems, Inc., of Bellingham, Washington, established a unique engine management system. The work is a spinoff from Langley's Engine Management and Predictive Analysis System (EMPAS) project. EMPAS strives for innovative approaches to providing detailed information on current engine status, as well as prediction of future engine states, including potential engine failure. EMPAS is seen as a form of predictive artificial intelligence. Pilots would be advised and provided possible corrective actions by EMPAS.

This system integrates newly developed and commercially available software, hardware, sensors, and display technologies—technologies produced or under development by Vision Microsystems, Inc.

The EMPAS effort has facilitated the company's creation of a product line of advanced engine monitoring systems. This includes the state-of-the-art VM1000 that combines ten individual instruments into one high-tech display. This engine instrumentation caution advisory system provides full sweep graphics, giving a pilot fast visual reference to operating limits and trends. The VM1000 automatically tracks changes in engine performance during flight. Microprocessor control, plus a flat panel, high-contrast display, support the firm's slogan: "Just one look...is all you need."

The VM1000 can be enhanced by the addition of the EC-100, an electronic checklist and cautionary system. Sporting a full color, easy-to-read alpha-numeric display, the EC-100 provides an array of information about engine and aircraft operating boundaries. The electronic checklist is designed around the operating characteristics of the pilot's aircraft. A number of emergency situation checklists are incorporated in the unit, such as "engine out," with data displayed at the pilot's fingertips in those moments when the cockpit workload is greatest. Checklist categories are arranged logically, making it easy and natural to select them.

Vision Microsystems also has produced the EPI 800, engine management instrumentation that consists of six individual $2^{1}/_{4}$-inch-diameter gauges. The EPI 800 has an onboard computer that continually analyzes engine operation and reports any abnormalities to the pilot. A built-in memory records data such as total engine hours and total fuel utilized.

Currently, the instruments are manufactured for experimental aircraft. However, the firm is in the process of obtaining approval for use in certified aircraft. The commercial potential of these engine management and predictive displays is great, offering the general aviation industry a cost-effective engine monitoring system heretofore only available, in part, to the airline industry. This technology is usable at all levels of general aviation, from the kit plane industry up to the business class, twin engine aircraft. Potential commercial uses are also foreseen in cruise ships, freighters, motor yachts, as well as private sector land vehicles, such as taxis, buses, and other revenue producing vehicles. ❖

Vision Microsystems, Inc.'s VM1000 combines instruments into one high-tech display.

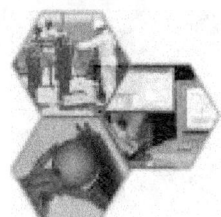

Taking a Position
Commercial Benefits—Spinoffs

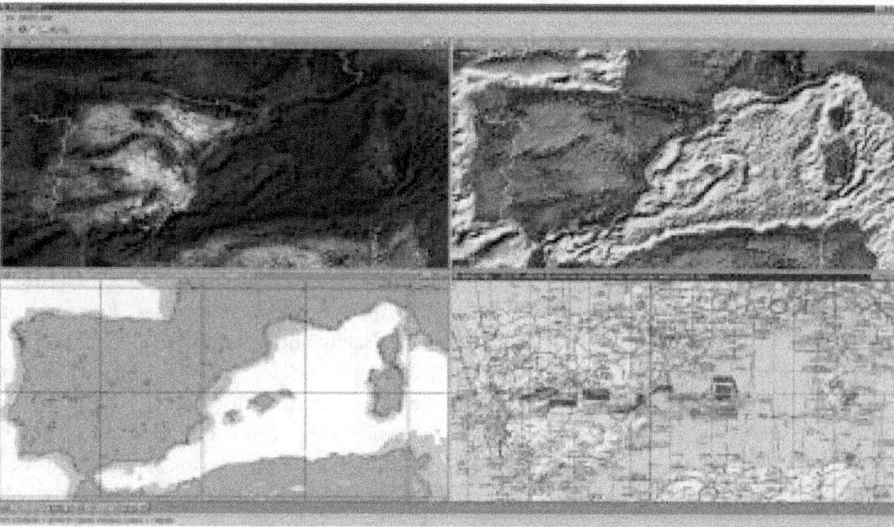

It can be a private pilot's worst nightmare: flying in a small plane, lost and surrounded by unknown terrain. Eliminating such a nail-biting worry is among the benefits offered by software programs made possible by mixing the talents of entrepreneurs and space scientists.

Dubbs & Severino, Inc., an Irvine, California-based firm, has created a virtual window on the world through its Position Integrity™ product line that includes TerrAvoid and Tactical Display. The products are the result of the Technology Affiliates Program at the Jet Propulsion Laboratory (JPL) in Pasadena, California.

Through this JPL activity, American industry gets a boost from NASA experts and also facilitates business use of intellectual property developed for the space program. Several years ago, the Technology Affiliates Program introduced the start-up firm to NASA's Dr. Nevin Bryant, who headed JPL's Cartographic Applications Group.

The JPL group had developed GeoTIFF, an architecture standard that provides geo-location tools for mapping applications. By the early 1980s, NASA had begun exploring the capability for precise positioning using the military's Global Positioning Satellites (GPS). Applications were many, including orbit determination for Earth-orbiting satellites and precise geodetic studies of crustal motion and plate tectonics. GeoTIFF is now in the public domain, and its use for commercial product development has evolved into an industry standard.

Under a licensing agreement from JPL, GeoTIFF proved an ideal tool for Dubbs & Severino, allowing the firm to move forward on plans to develop low-cost software packages. The JPL-designed architecture was adapted by the private company, giving them a jump-start on their products' specific attributes. "JPL gave us a demonstration and opened up the red carpet. It was a match made in heaven," says company president Bob Severino.

The idea for mapping software to help private airplane pilots was spawned, in part, by tragedy. A fatal crash had taken the life of a pilot friend of Severino, and technology, he believed, could have helped avert the accident.

The software packages that resulted were designed primarily for military sponsors and are now positioned for consumer market placement. The software can be run on a battery-powered laptop. TerrAvoid is a terrain avoidance system that graphically maps out and highlights threatening conditions in a pilot's flight area. A map scrolls along, as in a video game, presenting a sweeping 360-degrees of view. Red areas indicate the tallest mountains, while safe sections are depicted in green. Proportions change in real-time as the pilot moves through hilly terrain. The software integrates GPS data with maps sourced from a CD-ROM.

The Tactical Display is computer software that also co-registers real-time GPS data with local maps on CD-ROM. It is a moving map that details the exact position of the pilot.

Incorporating the unique features of JPL's GeoTIFF, this software can be adapted to operate with any map, chart, or photo image in the world. Four windows can be displayed at once, a useful function for an aviator who simultaneously can scan maps, charts, photo images, and sketches at different scales and zoom levels.

These two Dubbs & Severino products bring the same features available in military and commercial jets to general aviation pilots, for one-twentieth the cost. ❖

Position Integrity™ is a trademark of Dubbs & Severino, Inc.

Dubbs & Severino, Inc.'s Position Integrity™ software can display maps and other imagery side by side.

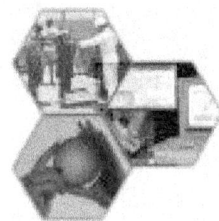

Navigating the Airways
Commercial Benefits—Spinoffs

The next-generation aviation navigation system is under development, a system that makes use of Global Positioning System (GPS) satellites for aircraft to accurately define their positions.

The U.S. Federal Aviation Administration (FAA) is implementing the Wide Area Augmentation System (WAAS). WAAS is the first to greatly enhance flight safety and airline efficiency, and decrease costs associated with ground-based navigation aids. The prime contractor for the system is now Raytheon Systems Company of Lexington, Massachusetts.

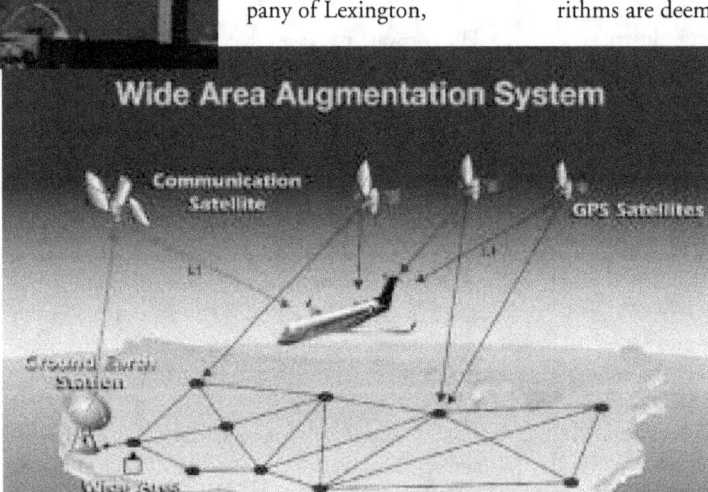

Courtesy of Jay Berkowitz, LAWA

WAAS is the first aviation navigation system to greatly enhance flight safety and airline efficiency.

Industry and government teams are working together to establish WAAS. One of the tools being utilized is software developed by NASA's Jet Propulsion Laboratory (JPL), located at the California Institute of Technology. JPL's software includes two components: Real-Time GIPSY (RTG, where GIPSY stands for GPS-Inferred Positioning System) and WADGPS Ionospheric Software (WIS, where WADGPS stands for Wide Area Differential GPS). In the case of WAAS, JPL developed state-of-the-art software for GPS applications, enabling precise orbit determination and estimation of GPS clock and atmospheric parameters—the key measurements that must be computed in real-time for operational WAAS. The precise estimators that are being installed into the FAA WAAS system are a great improvement over conventional civilian GPS accuracy and provide key integrity information in real-time, ensuring that the system provides multiple levels of safety, redundancy, and robustness for safety purposes.

The JPL software has been helpful in planning and implementing the WAAS system. The licensing program is transferring technology worth millions of dollars to the U.S. private sector from JPL. Together, these software packages are contributing to a navigation system that is expected to have a beneficial impact on the lives of millions of people, while keeping U.S. satellite tracking technology in a position of world leadership. RTG and WIS algorithms are deemed critical to the successful implementation of WAAS, a system that in the near future will provide all airliners in U.S. airspace with meter-level, real-time knowledge of position using GPS satellites.

Future WAAS benefits will include airplanes flying more direct routes, precision approaches into nearly any properly configured airport in the U.S., the gradual decommissioning of older navigation systems, which are expensive to maintain and operate, and reduced and simplified navigational equipment aboard aircraft. WAAS has the capability to provide aircraft flying in the U.S. a single system to navigate their way from takeoff to landing and from taxi to the terminal.

The 21st century airspace system will involve ground stations that receive, analyze, and refine signals from the GPS satellites and transmit the information via communication satellites to all aircraft flying within U.S. airspace. The WAAS ground stations will be located at some 35 air traffic control facilities, spread across the country.

It is anticipated that WAAS will save billions of dollars in fuel and aviation costs, while greatly enhancing aviation safety for millions of travelers. Additional systems similar to WAAS are also planned in other countries, using JPL's RTG software and WIS for precise real-time GPS corrections. ❖

A Tail Not to Get "Torqued" About

Commercial Benefits—Spinoffs

Helicopter owners and pilots now have an option to buy a device that improves performance by reducing the craft's natural tendency to turn due to torque.

The Langley Research Center granted a license for "tailboom strake" technology to Boundary Layer Research (BLR), Inc., of Everett, Washington. The Royal Airforce of New Zealand is currently using the NASA-patented device. The Australian Defense Force is installing the technology on their fleet of UH-1 helicopters. In the U.S., BLR is applying for Federal Aviation Administration (FAA) certification to make the technology available to civil operators and owners.

A NASA-Army team of researchers developed the tailboom strake device, a straightforward technology that redirects winds created by a helicopter's main rotor. Typically 10- to 15-feet long, the helicopter strakes run the length of the tailboom and extend diagonally from the surface about 3 inches. One strake is located near the top of the tailboom and the other near the bottom of the tailboom.

Reported benefits are improved stability, less horsepower needed for the tail rotor, improved side-to-side (yaw) control, enhanced altitude performance, increased payload capability at high altitude, reduced fatigue for tailboom and related flight critical components, and lower maintenance costs.

BLR is a small company that develops and markets performance aero-modifications for aftermarket application, selling them in a kit form with FAA approval. Company products include vortex generators, winglets, wing tanks, airframe strakes, stall strips, and landing gear upgrades.

In August 1997, a fortuitous Internet search using the word "strake" led company president Robert Desroche to a page detailing the NASA and Army work with helicopter tailboom strakes. The technology had been patented by the space agency in 1987. Desroche then contacted NASA and began working towards licensing the technology. A commercial license to market the NASA-patented strake technology was granted on December 16, 1998. The next day, the company sold its first product resulting from the technology.

The strake technology represents BLR's first venture into the rotorcraft modifications market. "We are very pleased to be selected by NASA to further develop this technology and we are anxious to get it to the operators where it can do some good. Operators work a very delicate profit margin and this technology will help tip the scale in their favor since it can reduce maintenance and increase performance. The fact that it improves safety by improving yaw control margins is just icing on the cake," says Desroche.

In February 1999, BLR announced delivery of its first Tailboom Strake Kit to the U.S. Army Aviation Center at Fort Rucker, Alabama. The Army is evaluating the device as part of an enhanced anti-torque system for its single rotor helicopters. That same month, the company also announced the placement of an advance order by a commercial company for 22 Tailboom Strake Kits, enough to modify an entire fleet of helicopters used in carrying sightseeing passengers over the Grand Canyon.

According to David George, BLR's vice president and manager of the newly formed rotorcraft division, the company has plans to design kits for 16 different helicopter models within the next 2 years. BLR's future tasks are to refine the technology, optimize the design for the different models of helicopters, and to secure FAA approvals for the modifications.

Desroche explains, "Helicopter operators can see that these two little pieces of technology have the power to save lives, and we are working as quickly as we can to make it available." ❖

Boundary Layer Research, Inc., is marketing NASA's patented "tailboom strake" technology, improving the safety of helicopters.

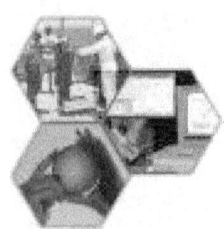

Here's a Switch

Commercial Benefits—Spinoffs

Coming face to face with a panel full of switches and dials in a small, general aviation aircraft cockpit is more than an eyeful. It can be dangerous as well.

Modern aircraft have a multitude of systems and controls blended into what is known as the instrument panel. Advancements in aviation technology have driven the panel's shape and size. But in many ways, such avionics improvements can make the instrument panel cluttered and harder to expand. More importantly, a jungle of devices and the addition of new systems can reduce the safety of flying conditions. The causes of numerous aviation accidents have been traced to unrecognized or improperly interpreted instrument readouts.

According to research, progressively deteriorating conditions, such as weather, instrument failure, and aircraft system malfunctions, are significant causes in general aviation accidents. In any of those cases, problems go unrecognized, are not understood, or are mishandled by the pilot. Problems predominantly begin as small symptoms that go undetected by the pilot, then evolve into larger problems, with potentially dangerous outcomes. Such problems could possibly be averted with an intelligent human factors-engineered pilot alerting system.

Through a NASA Advanced General Aviation Transport Experiments (AGATE) competitive award negotiated by the Langley Research Center, Mod Works, Inc., of Punta Gorda, Florida, has applied human factors engineering to the instrument panel to optimize human/airplane interaction. The result is the "Smart Panel," offered by Mod Works to reduce pilot workload and increase flight safety.

Mod Works has designed an instrument panel for installation in certified aircraft offering features geared toward simplification in cockpit design for human factors. The affordable panel is an easy retrofit into aircraft, updates older, confusing panels to present day standards, and reduces the number of

Mod Works, Inc., has designed an instrument panel that offers features geared toward simplification in cockpit design.

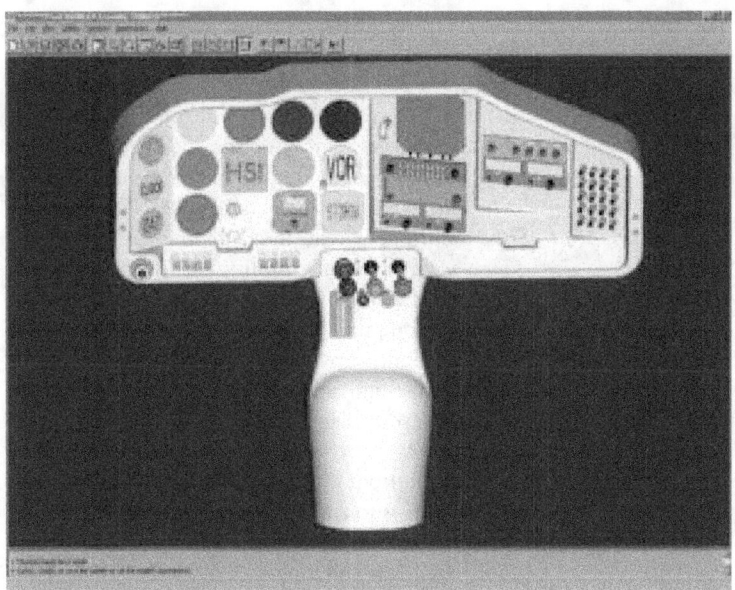

The company started work on its instrument panel concept by creating a 3-D computer model.

control panel instruments. Ergonomically designed, the panel reduces pilot workload and eyestrain, and is completely and quickly removable for repair or maintenance.

In addition, the Mod Works "Smart Panel" includes switches, circuit breakers, and the VM1000 engine monitoring system, developed by Vision Microsystems, Inc. (see p. 48). The VM1000 is considered by pilots to be the most advanced engine monitoring system available, where instruments are combined into one display.

Mod Works started work on the new panel concept by creating a 3-D computer model of the interior structure of a popular general aviation aircraft. A database was then established of instruments and avionics currently on the market.
By placing the desired equipment in the 3-D space of the interior model, engineers at the company were able to optimize location, viewing angle, hand access, and visibility. The information depicted on each instrument was categorized by necessity and function. Human factors engineering drove the panel design, rather than evolution or space availability.

From there, Mod Works transferred 3-D math data to fabricated instrument panels. After two years of evaluation and then certification, Mod Works introduced the aircraft panel for commercial sale.

The Mod Works cockpit design innovation integrates emerging and commercially available software and hardware technologies now in the marketplace or under development by avionics vendors. By using commercially available products, risk and cost of the general aviation end product can be controlled.

Mod Works is confident that a general aviation market exists for human factors-engineered cockpits. The need for reduced pilot workload and enhanced intuitive decision-making, tied to the human factors engineering know-how of Mod Works, makes for improved flight safety. This capability, the firm believes, has excellent market potential since no such product is available in the retrofit general aviation market. ❖

Park Smart

Commercial Benefits—Spinoffs

Merritt Systems, Inc., developed the Robot sensorSkin™ for NASA to enhance robots that work in constrained, hazardous, or dynamic environments.

Ever find yourself in pursuit of a parking spot? No doubt all motorists have at times found themselves searching level after garage level for the sanctity of an open space. Advanced sensor technology to improve the mobility of robots has been applied to precisely guide a driver to a free space.

The Parking Garage Automation System (PGAS) is based on Robot sensorSkin™, a technology advanced by Merritt Systems, Inc., (MSI) of Oviedo, Florida. This technology was developed under a NASA Small Business Innovation Research (SBIR) contract awarded by the John F. Kennedy Space Center. NASA needed to enhance robots that worked in constrained, hazardous, or dynamic environments—conditions found in Space Shuttle operations, for example.

The PGAS can be installed around and within public parking garages to monitor available parking spaces.

MSI teamed with NASA to create a system that contains smartSensor™ modules and flexible printed circuit board skin. The Robot sensorSkin™ was originally crafted to be cut to fit, shaped, and installed on a variety of robot manipulator arms to increase their use near critical flight hardware. As a result, robots could better detect and steer clear of obstacles in their path.

In one application, the sensorSkin™ technology was used on a NASA prototype payload inspection and processing automaton, a long-reach serpentine inspection robot. This preflight inspection and verification robot for Shuttle payloads was outfitted with a skin of non-contact proximity, ultrasonic, and infrared smartSensor™.

This state-of-the-art sensor system work was adapted and applied by MSI to various commercial applications. For the PGAS, this smartSensor™ network can be installed around and within public parking garages. Once in place, these sensors guide a motorist to an open facility, and once the driver has entered, to an available parking space. The sensors use noninvasive, reflective-ultrasonic technology for high accuracy, high reliability, and low maintenance. Employed as a network, the sensors are connected to a garage computer, which in turn, is tied to outdoor smart parking signs via radio-frequency modem links. These signs can first signal to customers what garages have available spaces. Inside the garage, additional smart routing signs mounted overhead or on poles in front of each row of parking spots guide the motorists to any open parking slots.

Many public parking garages employ devices that only count cars entering a lot, not exiting. Therefore, garage operations typically don't know how many locations are vacant at any given moment. However, the PGAS offers parking lane vehicle counting with accurate parking area occupancy reporting. Armed with such knowledge, garage operators can realize increased revenue by keeping parking areas completely utilized.

Thanks to its modular design, the PGAS can be quickly and easily installed at a modest cost, within a new or retrofit setting. It requires no structural modifications to existing concrete construction. An added security feature is a license plate recognition system. This function can automatically read entering and exiting vehicle license plate numbers, logging this information into a computer.

Ronald Remus, CEO/president of MSI, explains that the SBIR commercialization of the company's technology is proving to be extremely successful. MSI has participated in several major security and parking industry trade shows, showcasing their innovative sensor work. Significant interest has recently been raised by one major railroad group, eyeing the MSI technology to provide vehicle classification for rail cars. The purpose is to prevent accidents caused by having oversized rail cars traveling on the wrong track, resulting in accidents caused by the oversized rail cars colliding with bridges or tunnels. ❖

sensorSkin™ and smartSensor™ are trademarks of Merritt Systems, Inc.

Camera-on-a-Chip
Commercial Benefits—Spinoffs

The requirement for low-cost, compact imaging systems used in spacecraft has made cameras the size of a computer microchip possible. Applications of this state-of-the-art technology include personal computer video conferencing, digital still cameras, medical instruments, toys, and various automotive applications.

Photobit Corporation of Pasadena, California, first received exclusive license to a new type of image sensor developed at the Jet Propulsion Laboratory (JPL). The JPL-invented technology was the complementary metal-oxide semiconductor Active Pixel Sensor (CMOS-APS).

CMOS-APS technology enables the integration of a complete imaging system, including pixel array and control area, onto a single piece of silicon. One benefit is that it greatly reduces power consumption and lowers the number of parts needed in finished imaging products. Furthermore, by combining all camera functions—from the capture of photons to the output of digital bits—CMOS sensors offer enhanced reliability, facilitate miniaturization, and allow on-chip programming of frame size, exposure, and other parameters. Unlike conventional charge-coupled device (CCD) technology, CMOS sensors use the same manufacturing platform as most microprocessors and memory chips. Therefore, the CMOS devices are more cost-effective and easier to produce in comparison to CCDs.

Photobit, an entrepreneurial spinoff firm established in 1995 and based upon the JPL work, obtained the licensing rights to the CMOS-APS technology with the goal of furthering and marketing the revolutionary solid state image sensor. In early 1999, Photobit announced the issuance of a broad U.S. patent for camera-on-a-chip technology. The company's priority now is demonstrating the superior nature of the CMOS technology over CCD technology introduced in the 1970s. High-performance digital sensors that use CMOS architecture have been created at Photobit, "to set new performance standards for videoconferencing, digital still cameras, broadcast television, medical, agricultural, and children's applications," says Photobit's CEO and founder Sabrina Kemeny.

Two new products in Photobit's line of off-the-shelf videoconferencing chips were unveiled in 1999. The new sensors produce color or monochrome full-frame 8-bit digital video at 30 frames per second. They feature electronic pan, tilt, and zoom, auto-exposure (with manual override), and full programmability via a serial interface. Besides videoconferencing, the devices will be used in video cell phones and other small-format applications.

Photobit Corporation's camera-on-a-chip is the result of a license granted to use NASA's CMOS-APS technology.

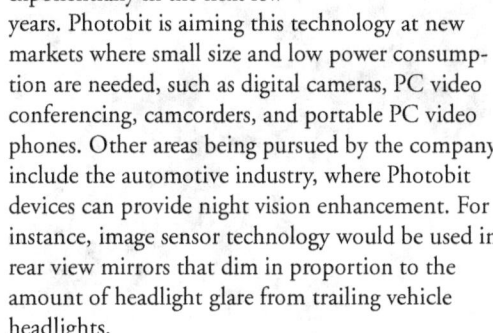

Photobit offers a unique choice of off-the-shelf and custom products. Officials at Photobit are confident about the commercial use of the technology, as image capture markets are expected to expand exponentially in the next few years. Photobit is aiming this technology at new markets where small size and low power consumption are needed, such as digital cameras, PC video conferencing, camcorders, and portable PC video phones. Other areas being pursued by the company include the automotive industry, where Photobit devices can provide night vision enhancement. For instance, image sensor technology would be used in rear view mirrors that dim in proportion to the amount of headlight glare from trailing vehicle headlights.

In the medical market, CMOS-APS technology can be tapped for x-ray products, including bone mineral density measurements. This allows a physician to track the onset of osteoporosis with less than one-hundredth the dosage of a dental x-ray to the patient.

NASA and the United States Space Foundation recognized Photobit's efforts in commercializing the compact imaging system by inducting Kemeny; Eric Fossum, chairman and chief scientist; Robert Nixon, deputy product division manager; Barmak Mansoorian, new-product marketing manager; and Roger Panicacci, senior engineer, into the Space Technology Hall of Fame. ❖

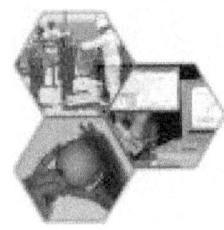

Performance Boosting Additive
Commercial Benefits—Spinoffs

An inexpensive way to increase the performance of air conditioners, heat pumps, refrigerators, and freezers has found a home in the commercial sector, spurred into existence by the need to thermally control NASA spacecraft.

Through Small Business Innovation Research (SBIR) funds from the Goddard Space Flight Center, Mainstream Engineering Corporation of Rockledge, Florida, developed a chemical/mechanical heat pump. The system makes use of environmentally acceptable working fluids, in particular, non-ozone-depleting substances.

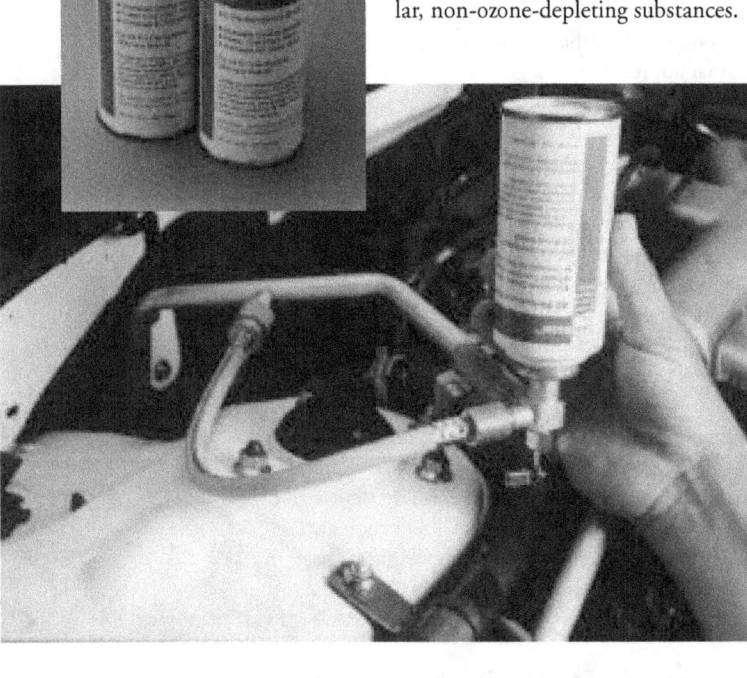

QwikBoost™, developed by Mainstream Engineering Corporation, is a refrigerant additive that increases cooling capacity.

As an indirect result of the SBIR-supported research, Mainstream Engineering has developed a unique, patented, low-cost refrigerant additive, called QwikBoost™. The product works by increasing the cooling capacity of the refrigerant. QwikBoost™ circulates through the refrigeration system in a manner similar to that of the lubricant. It has a high affinity for liquid hydrofluorocarbon and hydrochlorofluorocarbon refrigerants and also exhibits a significant heat of solution when mixed with them. This solution heat increases the available cooling capacity—the latent heat of the refrigerant during evaporation. Thus, the performance of the system is increased.

When production of ozone-depleting refrigerants was halted in the United States as part of the Environmental Protection Agency's Clean Air Act, affected units switched to another type of refrigerant. However, in doing so, these same units suffered a performance and efficiency reduction. Lower vapor-compression system efficiencies meant more electrical consumption resulting in the production of more power plant emissions.

QwikBoost™ has shown to be an environmentally safe fluid with zero ozone depletion potential, improving the performance of vapor-compression heat pumps, air conditioners, and refrigeration systems by as much as 20 percent. This results in a reduction in energy use and expenses for the equipment owners and a reduction in pollution generated by power plants.

QwikBoost™ is currently being marketed and sold in certain automotive air-conditioning systems and refrigeration units. The product is packaged in a 3-ounce can, pressurized with R-134a refrigerant. Once the additive is introduced into the system, it remains active for the life of the system and does not need to be replaced. An increase in automotive air-conditioning cooling capacity means faster car cool-downs and more cooling. This is a desirable attribute given the reduced capacities of new auto air-conditioning systems operating with or retrofitted to use R-134a refrigerant.

Testing of the product has shown that it will not adversely affect system lubrication or compressor life. Adding QwikBoost™ resulted in reduced wear properties compared to the lubricant alone. These tests also indicated that the product reduced the accumulation of wear metals in the lubricant.

Mainstream Engineering Corporation received the Tibbetts award from the Small Business Administration during an October 1997 White House ceremony for its commercialization of the performance enhancing additive technology.

Future use of QwikBoost™ appears bright, as residential and commercial air-conditioning and refrigeration systems will undoubtedly face regulations to reduce energy consumption. This additive offers a fast, easy, and inexpensive way to meet these future goals.

Meanwhile, new military and NASA units that take advantage of QwikBoost™ will be more efficient, smaller, and lighter—all desirable features for aircraft and spacecraft applications. ❖

QwikBoost™ is a trademark of Mainstream Engineering Corporation

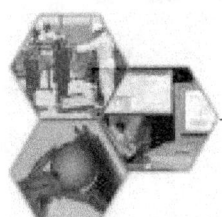

E-mail on the Move
Commercial Benefits—Spinoffs

Check your pocket...you've got mail! Technology originally developed at NASA now lets users send and receive electronic mail (e-mail) through regular or cellular telephones using a small, hand-held unit.

PocketScience™ Inc., of Santa Clara, California, is a privately held company founded in 1995, aimed at making mobile e-mail service not only affordable, but easy to use. The company is pursuing its mission by melding key technologies, expertise in consumer electronics and electronic messaging, and strategic partnerships. At its inception, PocketScience™ was a member of the NASA Ames Technology Commercialization Center.

A new hand-held device uses technology originally developed at NASA's Jet Propulsion Laboratory. Through its involvement with Ames' Technology Commercialization Center, PocketScience™ was able to use space probe communications technology and adapt it for advanced signal processing on Earth.

The idea behind the technology is to overcome the limitations of current mobile messaging solutions and provide access to e-mail anywhere. PocketScience™ created its first offering, the PocketMail® device. PocketMail® makes it possible to send and receive e-mail from anywhere in the world without turning on a computer. The device gives regular e-mail users a practical alternative to laptop computers and wireless devices when trying to send e-mail on the go. International travelers no longer have to be confronted with the nightmare of incompatible electric systems and telephone jacks while far from their home country. The mobile messaging device also permits the transmission of faxes.

The firm's device operates by holding it against a phone handset and pushing a button. Even under the harshest conditions, the small, 9-ounce portable unit, can send and receive e-mail through most phones worldwide, including cellular, cordless, Integrated Services Digital Network (ISDN), office and hotel PBX phones, and pay phones. In airport terminals, on busy street corners, and in other noisy locales, the device functions by incorporating special modulation schemes, error-correction, data compression, and data communications protocols. The technology is packaged to fit in a shirt pocket, runs on two AA batteries, and is about the size of a calculator. A modest monthly charge, beyond the initial cost of the electronic device, gives users access to e-mail services.

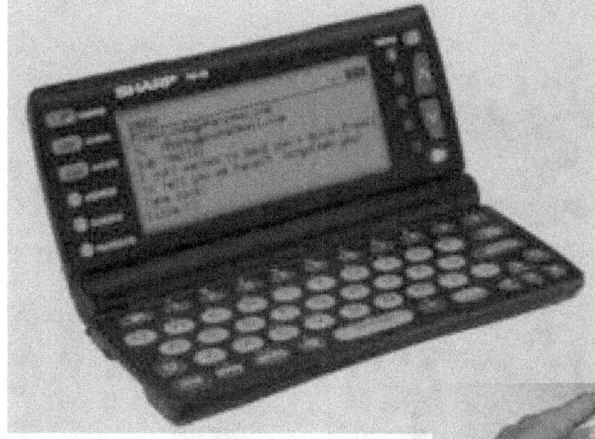

PocketMail® by PocketScience™ is a hand-held e-mail device that uses technology originally developed at NASA's Jet Propulsion Laboratory.

Users of the device compose their message, dial a nationwide toll-free access number, then push one button while holding the device against the telephone handset to send and receive messages. No cables or special connectors are needed. By employing burst packet communications, no lengthy log-on is needed. The entire process usually can be completed in less than a minute.

Electronic mail has been on a spectacular growth curve. Its phenomenal climb as a major communication tool is supported by research surveys. One study suggests the number of e-mail users in the United States alone is expected to grow from 75 million in 1998 to 135 million in 2001. And those users will transmit, in the U.S., as many as 500 billion messages in 2001. Another survey statistic is that Internet users check their e-mail at least once a day, a habit that is on the rise from previous years.

"With e-mail rapidly becoming a preferred way to stay in touch with friends, family, superiors, staff, or clients, it's only natural that people will also want to be able to access and respond to their e-mail while away from their homes and offices," states a corporate background paper developed by PocketScience™.

The first PocketMail®-enabled products have been announced by JVC and Sharp Electronics. Several other U.S. and international consumer electronics manufacturers are currently evaluating PocketMail® technology for inclusion in their next generation of products. ❖

PocketScience™ is a trademark of PocketScience, Inc.
PocketMail® is a registered trademark of PocketScience, Inc.

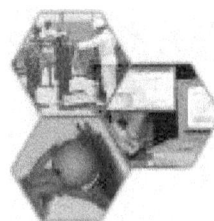

Ocean View
Commercial Benefits—Spinoffs

Analytical Spectral Devices, Inc.'s SeaSpec™ spectroradiometer was built to withstand the harsh marine environment.

Since August 1997, the OrbView-2 satellite has circled the Earth, providing never-before-seen color images of our planet's ocean and land surfaces. The on-board NASA Sea-viewing Wide Field-of-view Sensor (SeaWiFS) is taking these images. The images not only have commercial applications, but scientific researchers around the globe are using them to assess global warming and the Earth's complex biosphere.

A primary job for SeaWiFS is to provide quantitative data on global ocean bio-optical properties to the science community. This satellite sensor is shedding light on the primary productivity of the upper oceans and the fluxes of carbon dioxide and other trace gases across the sea-air interface.

In order to validate SeaWiFS and develop algorithms, Analytical Spectral Devices (ASD), Inc., of Boulder, Colorado, designed the SeaSpec™ spectroradiometer, an underwater system that is autonomous and rated to a submersible depth of 200 meters. SeaSpec™ instruments were developed with NASA Small Business Innovation Research (SBIR) funds through the Goddard Space Flight Center.

Unlike band-pass radiometers that supply just a few spectral bands, the underwater device has two charge coupled device (CCD) spectrometers to provide continuous spectra, from 350 nautical miles to 950 nautical miles. This allows the user to model information received from SeaWiFS, as well as all existing and future satellite sensors. The instrument itself is only 36 inches in length, with its outside diameter of 8 inches being an aluminum pressure housing. The instrument's autonomous operation and power management allow deployment from a float or buoy, which enables the sensor to avoid errors caused by light reflection and shadows from a ship's hull. The spectroradiometer was built to withstand the harsh marine environment.

ASD has been able to commercialize the engine of the spectroradiometer. The CCD spectroradiometers, resulting from the work on SeaSpec™, have been used to create two commercial products. Both now enjoy popular use for coastal research. One product, the FieldSpec® VNIR Dual CCD, allows for a small suitcase-sized spectrometer to be placed in a boat. An attached fiber optic cable can then be lowered 30 feet below the surface, allowing researchers to perform the same studies as SeaSpec™, although not at deep-water depths.

The second product, FieldSpec® VNIR-CCD spectroradiometer, provides the increased sensitivity demanded by many oceanographic applications. With the addition of fiber optic extension cables, this instrument can be used to measure water surface reflected radiance, down-welling solar irradiance, and in-water up- and down-welling infrared radiance to depths of 33 feet to over 65 feet.

Data collected by such ocean placed sensors are important to the planet's future. Solar irradiance measurements, for example, are helpful in understanding Earth's climate system. It is very important to establish and maintain a long-term record of this energy source.

Since 1990, ASD has established itself as a leader in portable spectroscopic measurement instrumentation. A variety of applications are served by ASD equipment, such as portable, visible, and shortwave infrared spectroradiometers for the environmental remote-sensing marketplace. ASD's instruments are in wide use for geology, ecology, agriculture, and marine and coastal oceanography. In this regard, ASD's instruments are exceptional for ocean color, water quality, primary production, underwater visibility, algal bloom characteristics, and sea floor optical characteristics. ❖

SeaSpec™ is a trademark of Analytical Spectral Devices, Inc.
FieldSpec® is a registered trademark of Analytical Spectral Devices, Inc.

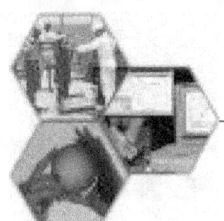

Long-Life Catalyst
Commercial Benefits—Spinoffs

Finding out which way the wind blows has stimulated development of a catalyst capable of converting toxic carbon monoxide to a nontoxic gas.

Langley Research Center scientists have developed low-temperature carbon monoxide oxidation catalysts. The requirement for these catalysts was driven by the need for recycling carbon monoxide and converting it back to carbon dioxide during the operation of closed-cycle carbon dioxide lasers in space environments. These catalysts were developed for a long-life, highly power-pulsed carbon dioxide laser, to be incorporated in a Laser Atmospheric Wind Sounder (LAWS) satellite. The catalysts were used to measure wind velocity on a worldwide basis.

The problem faced by space engineers and scientists was that the electrical discharges that energize such lasers generally decompose some of the carbon dioxide to carbon monoxide, resulting in a loss of laser power. The most practical solution to the problem is regeneration by catalytic recombination. To minimize energy consumption, recombination should be done at ambient laser temperatures with no addition of energy to the catalyst or laser. What was needed was a catalyst that met these properties, along with the ability to remove carbon monoxide and formaldehyde from the air in enclosed spaces.

Scientists and engineers at STC Catalysts, Inc. (SCI), of Hampton, Virginia, supported the Langley development of the catalyst, and are co-inventors on the patents. The firm has an exclusive license from Langley to manufacture carbon monoxide oxidation catalysts for use in carbon dioxide laser applications. The catalyst permits the closed cycle operation of a laser for billions of pulses without replenishing the operating gases. Also, the catalyst prolongs laser life, reduces power output fluctuations, and can be customized to fit any laser.

SCI, a subsidiary of the STC Group, Inc., manufactures the noble metal reducible oxide catalyst, consisting primarily of platinum and tin oxide deposited on a ceramic substrate. It is an ambient temperature oxidation catalyst that was developed primarily for use in carbon dioxide lasers.

SCI has an exclusive license to manufacture and distribute the catalyst for all laser applications. The firm is also furnishing the catalyst for other applications through additional agreements with NASA and with the Rochester Gas and Electric Company in Rochester, New York, who holds a license for controlling air quality in inhabited spaces.

Energy conservation and indoor air quality are important but often conflicting priorities for gas and electric utilities and their consumers. One widely used method of conserving energy is to reduce the exchange of indoor and outdoor air by tightly sealing buildings. But such construction can result in the significant buildup of contaminant gases, requiring efficient methods of removal.

Carbon monoxide buildup, in particular, is dangerous in inhabited spaces. In some instances, it can be released in lethal doses by faulty furnaces or poorly ventilated fireplaces. While carbon monoxide alarms have been developed, a preferable solution is to remove carbon monoxide continuously and rapidly.

Now, thanks to the room temperature catalyst, toxic gas can be oxidized to nontoxic carbon dioxide when placed in air-conditioning systems. An additional benefit of the catalyst is that it also removes formaldehyde from air by oxidizing it to carbon dioxide and water. ❖

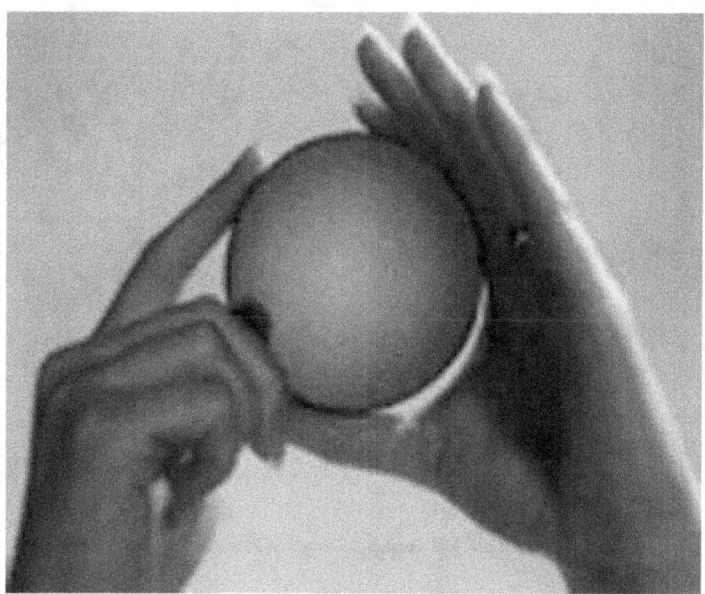

STC Catalysts, Inc.'s catalyst has the ability to remove carbon monoxide and formaldehyde from the air in enclosed spaces.

Marching Microbes

Commercial Benefits—Spinoffs

Biological products for a cleaner and safer environment are the business of Micro-Bac International, Inc., of Round Rock, Texas. The company has benefited from NASA Small Business Innovation Research (SBIR) funding by the Marshall Space Flight Center. NASA had contracted the group to develop a phototropic cell for water purification systems. This type of technology is necessary for use aboard space stations, and eventually, human habitats on the Moon and Mars. The phototropic cell requires less energy than competitive water treatment cells.

Since the project's completion in 1995, the company has been marketing a liquid purification product called Mega-BacTF®. This creation is currently manufactured and marketed for septic systems and wastewater ponds, lakes, and lagoons to degrade fat, oil, fecal matter, and other biologically derived wastes.

Mega-BacTF® is an all natural, non-pathogenic, environmentally friendly bacterial solution requiring no special clothing or equipment for treatment. Other treatment processes require the use of goggles, gloves, and laboratory coats. The metabolic process produce the enzymes necessary for degrading septic wastes and reducing odor. Sunlight is the energy source for the process, with no carbon monoxide or hydrogen sulfide generated.

The light-driven waste remediation system, using phototrophic bacteria, offers significant advantages over conventional aerobic or anaerobic waste treatment systems from the viewpoint of safety, flexibility, cost effectiveness, and performance. In geographical areas with adequate amounts of sunlight, this system is considerably more cost effective than traditional sludge aerobic systems.

Given increasing populations and the continual growth of industries, waste can overwhelm the environment. In the area of wastewater treatment, where each situation can vary, the firm has made a major contribution. Micro-Bac® is a product that takes on a multitude of organic wastes from food, humans, animals, and plants. In large wastewater systems, ponds, and lagoons, for example, the company's product improves operating efficiency by reducing sludge and grease. Furthermore, shock-loading problems are alleviated and odors are controlled. Micro-Bac® has been found ideal for municipal wastewater systems and other large septic systems. The product's naturally occurring microorganisms digest organic wastes and reduce biochemical oxygen demand, chemical oxygen demand, and the contaminant levels of total suspended solids.

Other products sold by the company are targeted at food processing and food service applications. For use at dairy plants, for instance, a product is offered to accelerate the natural degradation of wastes, such as fat, oil, milk, carbohydrates, and protein.

By harnessing the power of bacterial activity, petroleum production can be improved. With regular treatment, another company product can increase oilfield production, lower operating costs, reduce downtime, and improve physical oil characteristics.

Micro-Bac International has blossomed into a full-service research and development company specializing in manufacturing environmentally friendly biological solutions for a customer's problem. Sales of products are on the increase. Indeed, as a customer-driven company, Micro-Bac International's products are developed specifically for a client's individual environmental issue. Following extensive research and testing of a customer's sample, Micro-Bac International produces the microbial product that offers the most beneficial results. ❖

Treatment of wastewater utilizing the metabolic activity of light-sensitive bacteria was spurred by NASA's need for water purification processes useful in space habitats.

Mega-BacTF® and Micro-Bac® are registered trademarks of Micro-Bac International, Inc.

Fuel Cells for Society
Commercial Benefits—Spinoffs

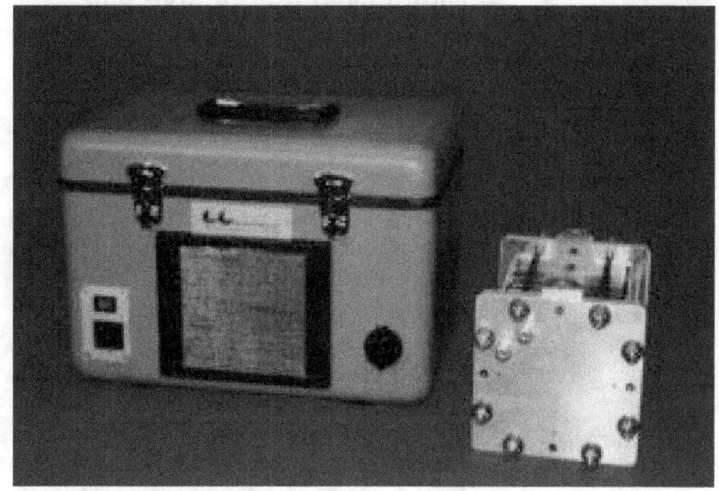

A new millennium technology headed for wider application in cars, homes, and power plants is the fuel cell. Operating like a battery, but never in need of recharging, fuel cells run on hydrogen and oxygen. The heart of a fuel cell consists of an electrolyte sandwiched between two electrodes. An electrochemical reaction between the gases produces electrical energy, water, and heat. The cell's waste product is water vapor, which can be used again in the system when broken down into hydrogen and oxygen. Clean running, efficient fuel cells are considered ideal for an energy-hungry world that is also endeavoring to reverse worrisome environmental issues.

ElectroChem, Inc., in Woburn, Massachusetts, has focused its engineering energies on solid polymer, or proton exchange membrane (PEM), fuel cells. The high-tech firm has sold hydrogen-oxygen fuel cells to the commercial market as a result of research performed under NASA's Small Business Innovation Research (SBIR) program at the Glenn Research Center.

Glenn and the NASA Wallops Flight Facility needed a hydrogen-oxygen PEM fuel cell for use aboard balloon-carried science platforms. The balloon-toted power system of choice is based on ElectroChem's PEM fuel cell technology sponsored by NASA SBIR funds.

NASA needed a high-power, long-duration power system to energize electronic equipment. Moreover, the power system had to deliver power continuously, in a safe, reliable manner, and under a wide variance in temperature and pressure. ElectroChem met the challenge. The final result was a working hydrogen-oxygen fuel cell stack producing 500 watts of power. Waste heat from the fuel cell maintained proper operating temperatures for the fuel cell, electronic equipment, and product water.

Similar to NASA's scientific balloons, business at ElectroChem has also soared skyward. The firm has gone from being solely a research company to commercializing products resulting in revenues of around $750,000 a year. Sales of its "fuel cell in a suitcase" have blossomed, with various government agencies and private industries, both in the United States and abroad, purchasing the energy-providing units. ElectroChem has started marketing its latest product, the EC-PowerPak200, a fan-cooled 200-watt fuel cell power system with automatic water draining and AC and DC outlets. Customers can plug an assortment of appliances, such as radios, into the fuel cell outlets.

Other uses for the PEM fuel cell include recreational vehicles, stand-alone regenerative power systems, rural electrification, and using it as a power source for underwater vehicles. ElectroChem is committed to further fuel cell research, aimed at reducing fuel cell prices to a range that residential homeowners can afford.

Recent studies of the fuel cell market suggest sales reaching $1.3 billion by 2003. ElectroChem's main goal is to develop small fuel cell power systems up to 5 kilowatts. This would allow the company to enter niche markets, then expand to cover off-grid applications in developing countries. To further critical research, the firm has established fuel cell laboratories in the United States and abroad.

"ElectroChem envisions a future where a sustainable global economy resides in harmony with a clean, healthful environment. Its mission is to provide fuel cell technology and products that this society will require for its energy supply," says Radha Jalan, CEO of ElectroChem. ❖

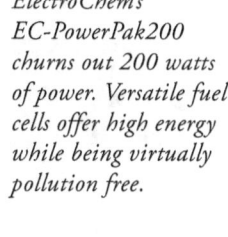

ElectroChem's EC-PowerPak200 churns out 200 watts of power. Versatile fuel cells offer high energy while being virtually pollution free.

Fuel cell portability and power makes ElectroChem's products viable for an assortment of uses around the world.

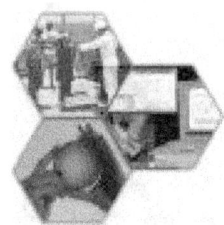

Making Visible the Invisible

Commercial Benefits—Spinoffs

Hydrogen is a clean-burning gas. Therefore, its flame is nearly invisible during daylight hours. A new system allows technicians to visually spot the invisible, making it possible to shut down the hydrogen source and extinguish a hydrogen fire.

Through NASA Small Business Innovation Research (SBIR) funds at the Stennis Space Center, Duncan Technologies, Inc. (DTI), of Auburn, California, has developed an infrared imaging system that provides color images of invisible hydrogen fires.

Older black-and-white infrared systems have been in use to detect hydrogen fires. However, this type of instrument has a tough time contrasting hydrogen fires from the surrounding background. Often the flame is displayed larger than it actually is, making it less useful in determining which nearby equipment and systems might actually be in danger.

At Stennis, the need for better tools to capture invisible hydrogen fires is mandatory. This NASA field center is the world's largest user of liquid hydrogen, consuming and storing great amounts of the propellant for testing Space Shuttle Main Engines, as well as other rocket propulsion systems and components.

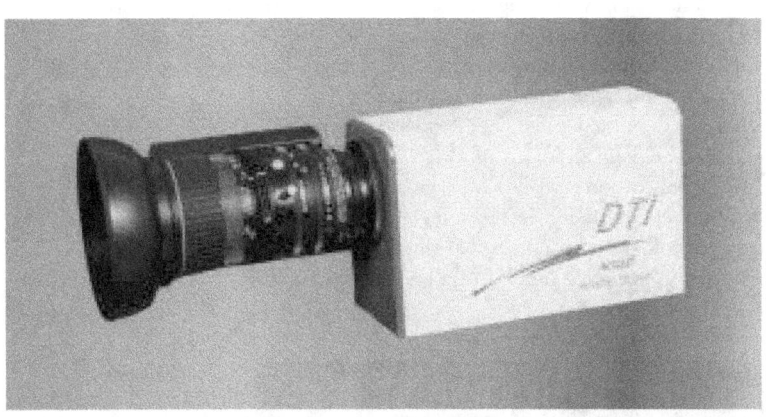

Duncan Technologies, Inc.'s HC1000 uses three different sensors to visually detect a hydrogen flame.

DTI was aware of research carried out at Stennis utilizing images in the infrared and near infrared to view hydrogen fires. DTI's proficiency in advanced imaging and electro-optical services gave the firm an upper hand in identifying the promise of alternate approaches for a hydrogen fire detection system. The infrared sensor technology is very expensive, resolution is poor, and because infrared sensors detect thermal phenomena rather than visible images, the resulting image can be difficult to interpret. By way of an SBIR award, the company looked to perhaps enhance and improve this type of imaging.

DTI built the HC1000, a system that uses three different sensors to visually detect the flame. Two of the sensors operate in the near infrared, while the other sensor works in the visible portion of the electromagnetic spectrum. Of the two near-infrared sensors, one is centered at a spectral band that detects strong water vapor emissions from the flame, while the other is at a band that detects minimal emissions to measure just the background image.

The flame is detected by subtracting the background image from the flame image and filtering the result. This process proved successful. A hydrogen flame can be isolated, then superimposed onto a color video image. A hydrogen flame of just 3 inches is detectable from 150 feet away.

The hydrogen camera was sold to NASA for use at Stennis. In addition, the system was sold to a private automaker to use on test stands, where work on hydrogen-based automobile engines is performed.

DTI developed a marketing plan to commercialize the multispectral imaging camera developed for Stennis. Presently, DTI is introducing a full line of multispectral cameras that enable new capabilities in machine vision, remote sensing, and surveillance applications. The camera products have generated worldwide interest for use in food processing, lumber grading, and quality control of pharmaceuticals. High quality color-infrared imaging for precision agriculture is yet another application.

"All of the multispectral cameras we are currently marketing are a direct spinoff of the hydrogen fire project," says Judy Duncan, DTI vice president. While the market for hydrogen fire detection equipment proved too narrow to support a commercial venture, Duncan adds that by generalizing the technology for industrial and agricultural applications, the marketing of advanced imaging technologies has proven quite successful.

"We invested company funds into continued development and refinement of the system and came up with the current line of products," Duncan says. "At present, we have many potential customers waiting for acquisition of an evaluation unit, which will hopefully be followed by order placement," she concludes. ❖

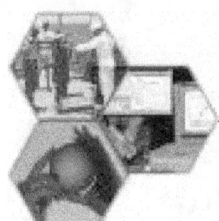

Laser Looking at Earth
Commercial Benefits—Spinoffs

Three-dimensional mapping of the Earth can help plan highways, survey areas for the laying of pipelines and utilities, and give city planners flood plain assessments to manage urban sprawl. Affordable and high-quality Earth surface data on demand is offered by TerraPoint™, LLC, of The Woodlands, Texas.

TerraPoint™, a business unit of Transamerica Real Estate Information Companies (TREIC), combines the business of providing topographic mapping services with the capability of NASA laser technology.

TerraPoint™ is marketing laser-generated digital, topographic data that can be obtained in a range of day/night, weather, and vegetation conditions. At the heart of this commercial service is laser technology sponsored by Goddard Space Flight Center's Technology Commercialization Office. This transfer of technology is the culmination of over 20 years of NASA research activities.

TerraPoint™ is headquartered on the Houston Advanced Research Center (HARC) campus in The Woodlands, Texas. The company has exclusive access to patented NASA optical technology for terrain mapping applications. TerraPoint™ is also integrating this technology into new airborne sensor designs, further improving the quality of the terrain products while continuing to drive down costs.

TREIC is a division of Fortune 500 Transamerica Corporation, one of the nation's largest financial services companies. Comprising four business units and headquartered in Dallas, Texas, TREIC provides property tax services, flood zone determinations, and detailed information on residential properties. With TerraPoint™ added as a TREIC business unit, Earth surface data on demand is now offered to customers.

This fusion of application and capabilities, business savvy, and space technologies also culminated in a melding of laser ranging and Global Positioning System (GPS) satellite hardware, replete with mapping software—all brought together in a miniaturized package that can be mounted in a light aircraft.

TerraPoint™'s Light Detection and Ranging (LIDAR) System capability has already been utilized in the largest high-resolution digital terrain mapping effort, scanning more than 1,700 square miles of Harris County, Texas. Harris County includes the nation's fourth biggest city, Houston. The result was over 275,000 precise terrain data points per square mile, with a vertical accuracy on the order of one foot.

TerraPoint™ has exclusive access to patented NASA optical technology for terrain mapping applications.

TerraPoint™ is committed to flying over the major U.S. metropolitan areas, and will offer off-the-shelf digital terrain data at prices meeting the budgetary needs of the industry.

The potential commercial customer base is extensive, given high-density and fast-turnaround data of the Earth's surface that are produced in a cost-effective manner. By introducing laser mapping, TerraPoint™ offers 3-D data that are ideal for charting flood plains, surveying pipeline routes, conducting highway design and planning simulations, placing communication antennas in a city, conducting shoreline and erosion surveys, acquiring river cross sections for hydrologic modeling, and assessing forest habitats.

Topographic data generated using laser technology are superior to those provided by traditional technologies, specifically the more commonly used radar and photographic technologies, says Dan Cotter, president of business development for TerraPoint™. "The technology permits 3-D mapping of cities, something not considered cost effective, that is until the introduction of TerraPoint™," Cotter adds. ❖

TerraPoint™ is a trademark of TerraPoint, LLC

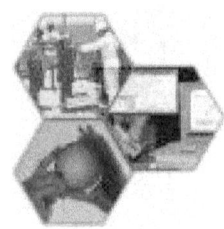

Ocean in Motion

Commercial Benefits—Spinoffs

The Earth is awash in oceans. Approximately 70 percent of the Earth is covered by water, largely in the form of four major oceans. In order for scientists to accurately take the pulse of Earth and ascertain the overall health of the globe and its precious biosphere, the use of satellites is critical.

Thanks to a special sea-viewing spacecraft, a daily snapshot of the world's oceans is now possible. Not only do oceanographers have a new tool at their disposal, but data taken by the satellite are available for a variety of commercial purposes.

Through a unique 5-year data purchase agreement between Goddard Space Flight Center and Orbital Sciences Corporation of Dulles, Virginia, vital information about the world's oceans is being gleaned from the vantage point of space. This government/private sector cooperation led to the

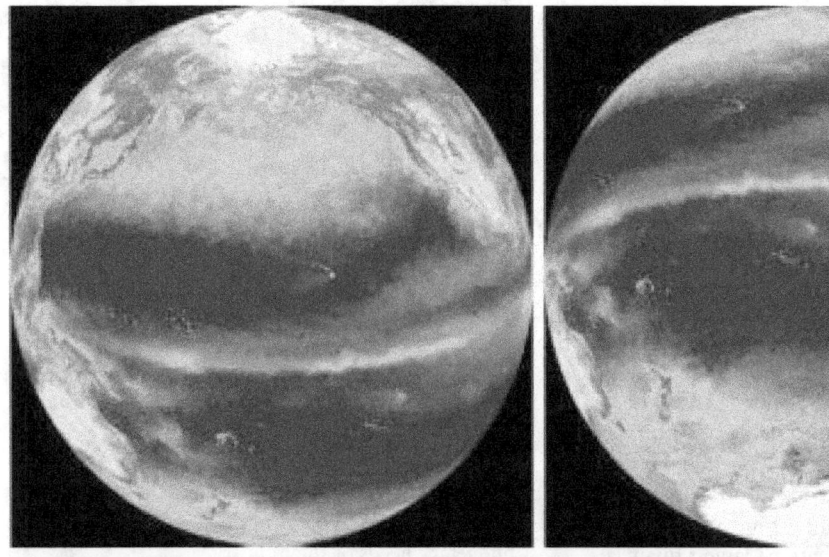

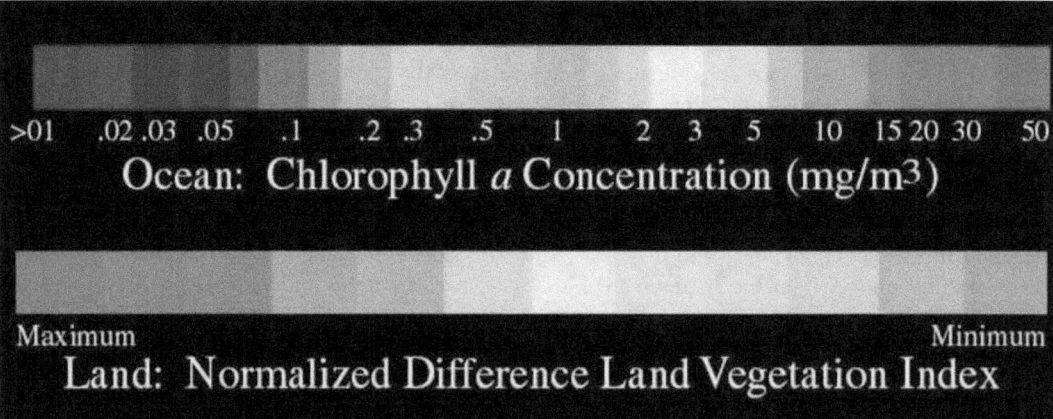

With Orbital Sciences Corporation's satellite-based, ocean-color imaging data, fishing fleets can locate areas where fish congregate.

64   Environment and Resources Management

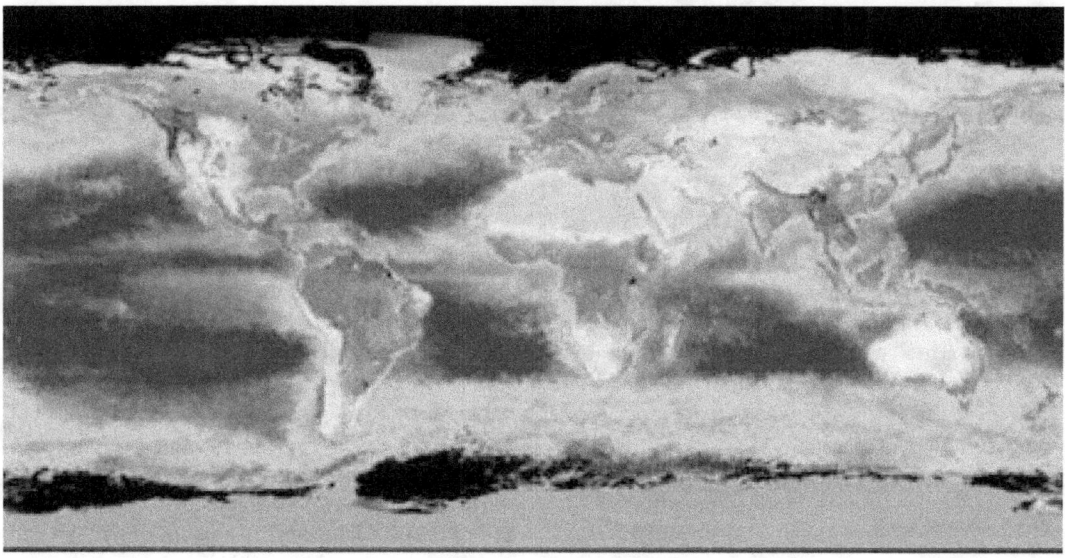

launching of the innovative Sea-viewing Wide-Field-of-view Sensor (SeaWiFS).

The satellite-based, ocean-color imaging instrument gathers data to analyze the process and rate by which the ocean cleans the atmosphere of carbon-based greenhouse gases through photosynthesis in phytoplankton on the ocean's surface. Such information is one piece of the puzzle of studying the impact of global warming. Ocean-color information is provided by the satellite each day. SeaWiFS is a multispectral sensor designed to distinguish between subtle color variations on the Earth's surface. This capability was first demonstrated aboard NASA's Nimbus 7 spacecraft, which is no longer operating.

Launched by Orbital Sciences on August 1, 1997, the SeaWiFS sensor first collected test images over the United States, Canada, and the Atlantic Ocean, transmitting these images to the main control center.

Through the innovative contract between NASA and Orbital, NASA is buying space-based ocean-color imagery produced by SeaWiFS for a period of five years. Although originally designed to observe the oceans, the sensor has also proven useful to study the land and atmospheric processes as well, explains Goddard oceanographer Gene Feldman. "As a result, we can monitor changes in the global biosphere with a single sensor over land and ocean," he adds.

After a year of continuous operations, SeaWiFS has yielded new insights into the impact of the El Niño climate anomaly on ocean life, fires in Florida, Mexico, Canada, Indonesia, and Russia, floods in China, dust storms in the Sahara and Gobi Deserts, and the progress of hurricanes, such as Bonnie and Danielle.

For Orbital Sciences and its Orbimage group, SeaWiFS data are being packaged for commercial sale. Satellite imagery is being marketed to organizations involved in environmental monitoring, high seas fishing, forestry assessment, and agricultural applications.

For example, take the satellite sensor's ability to distinguish phytoplankton-rich ocean regions from the clear ocean regions. Those data are used to create daily fish finding maps, allowing fishing fleets to focus on locations where many commercially important surface feeding fish, like tuna and swordfish, congregate. Imagery of phytoplankton and sediment in the ocean environment is valuable for pollution monitoring, "red tide" tracking, and global change studies.

In agriculture and forestry, the multi-spectral images offer an alternative to direct on-site inspection or expensive aerial photography. The accuracy of global crop yield forecasts can be increased, which is important for trading agriculture commodities. ❖

Thinking Like a Human
Commercial Benefits—Spinoffs

Among the benefits of neural networks are their ability to learn from experience and to generalize from their data set.

The interconnectedness of neurons in the human brain was the model in developing an advanced computer processor that can perform up to a billion computations per second.

Small Business Innovation Research (SBIR) contracts from NASA and other government agencies went into the development of a unique computer technology that "learns" as it goes. This technology is known as neural networking. A neural network is a class of computational methods that loosely imitate the function of the brain. Among the benefits of neural networks are their ability to learn from experience and to generalize from their data set. They are also fault tolerant and can exploit parallel systems for rapid processing.

By combining the funding from these SBIR contracts, Accurate Automation Corporation (AAC) of Chattanooga, Tennessee, developed the only Multiple-Instruction/Multiple-Data (MIMD) neural network processor on the market. The product is called the NNP® and is utilized in a number of computational applications.

The NNP® is used onboard the NASA- and U.S. Air Force-sponsored LoFLYTE® aircraft. LoFLYTE® has been built by AAC to demonstrate a computerized flight control system that learns as it flies—especially important for the demands of ultra high-speed flight. LoFLYTE® is being used to explore new flight control techniques involving neural networks, which are able to continually alter the aircraft's control laws in order to optimize flight performance and take a pilot's responses into consideration.

Over time, the neural network system could be trained to control the aircraft. The use of neural networks in flight would help pilots of future aircraft fly in quick-decision situations and help damaged aircraft land safely, even when the controls are partially disabled.

The NNP® consists of many interconnected control systems, or nodes, similar to neurons in the brain. Each node assigns a value to the input from each of its counterparts. As these values are changed, the network can adjust the way it responds. The NNP®'s extensive layers of nodes allow for high-speed parallel processing. The device is capable of implementing multiple neural network algorithms and paradigms. The NNP® comes in a number of versions, accommodating different computer platforms. The processors are used in many commercial and non-commercial problem-solving applications, including flight controllers, financial predictions, intelligent modems, medical image classification, on-line learning, robot control, satellite communications, and fault diagnosis and prediction.

An interesting example of how the NNP® functions is demonstrated by the LoFLYTE® aircraft. During normal flight, the neural controller will use the data it receives from the telemetry system to compute the most efficient flight characteristics and adjust the control surfaces accordingly. However, the neural control system has an enormous advantage over traditional control systems during abnormal and unexpected flight conditions. For example, if the control system determines that the rudder is not responding, it will adjust quickly and automatically to control the aircraft using the remaining flight surfaces. Neural network control is necessary in hypersonic vehicles, such as LoFLYTE®, where the center of gravity of the vehicle can change significantly throughout the flight. The neural network can adjust to changing flight conditions faster than a human pilot, greatly enhancing the safety of the aircraft.

Computer processing systems that are modeled after connections in the human brain are creating exciting changes in the way we do and will expect networks to function. ❖

Accurate Automation Corporation's neural processor, the NNP®, was developed using the interconnectedness of the human brain as a model.

NNP® and LoFLYTE® are registered trademarks of Accurate Automation Corporation

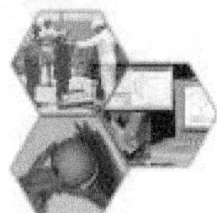

Early Warning Signs
Commercial Benefits—Spinoffs

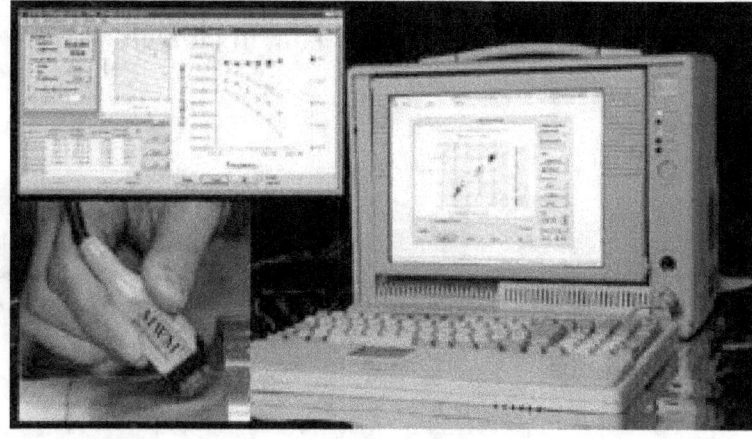

The tiniest cracks may spell big trouble for aging aircraft. Fatigue damage of sufficient size and density can weaken metal, leading to possible failure of a structure.

Through Small Business Innovation Research (SBIR) contracts with NASA's Goddard Space Flight Center, JENTEK® Sensors, Inc., of Watertown, Massachusetts, is now offering a system for nondestructive material evaluation, including thermal spray coating for characterization of porosity and thickness.

JENTEK® developed the GridStation™ Measurement System, based in part on a NASA need for characterizing anodic coating thickness, spacecraft contamination, and thermal barrier coatings. The innovation can provide age degradation monitoring, including fatigue, corrosion, and thermal aging in numerous materials, such as ceramics, composites, and metals.

JENTEK® markets two devices. The first, called the Meandering Winding Magnetometer (MWM™), is used for conducting, as well as for magnetic media. The other, known as the InterDigital Electrode Dielectrometer (IDED), is used for relatively insulating media. With the MWM™, JENTEK®'s first commercialized sensor product, magnetic fields are used to inspect conducting materials, such as metals, both magnetic and nonmagnetic. The MWM™ can spot miniscule cracks down to one- to two-millionths of an inch in depth. By comparison, conventional devices that use eddy current cannot generally find cracks of that size. Indeed, this capability is essential because clusters of microscopic cracks can affect the service life of a structure.

To reveal the microscopic cracks, the MWM™ signals are induced by specially configured conformable sensor arrays. These arrays minimize the requirement for precise positioning of the device and permit mounting the sensors permanently at any critical location for continuous monitoring of fatigue. Also, because this advanced device uses conformable sensors, it can inspect not only flat surfaces, but also convex, concave, and conical surfaces. Accurate property determinations can be obtained, regardless of the shape of the part being tested. For some applications, sensor scanning can be done at a fast-paced rate, up to 2 feet per second. An initial calibration of the sensor system in the laboratory permits the measurement of a wide variety of materials, without requiring the operator to use any standards or reference parts. JENTEK® has rigorously tested the system for performing very early stage crack detection in stainless steel, aluminum, and other materials.

Used with the company's product line of sensors, JENTEK®'s GridStation™ is a fully integrated and portable nondestructive property measurement system. Current and potential applications for the JENTEK® MWM™ and IDED sensors include characterization of coatings, fatigue damage mapping, crack detection and sizing, applied and residual stress measurement on ferromagnetic materials, object detection, material identification, and cure monitoring. On-line fatigue monitoring can be done in difficult-to-access locations on complex structures, such as aircraft, bridges, and heavy manufacturing equipment. The software environment uses measurement grid modules to convert impedance measurements to property estimates in real-time without requiring user interpretation. The data are stored in a "Grid Library."

While Goddard first considered use of the innovation for its spacecraft building and validation work, the Glenn Research Center capitalized the innovation and utilized the system to characterize ceramic thermal barrier coatings for turbine blade applications. The thermal barrier coatings allow operation of the turbine blades at higher temperatures. Monitoring the properties and degradation of the ceramic coatings, the metallic bond coat, and the super alloy blade material would provide for preventative maintenance, as opposed to costly unscheduled repairs when failures occur.

Neil Goldfine, JENTEK®'s president, says the NASA SBIR contract work helped advance the company's product into the commercial market. JENTEK® has sold and delivered GridStation™ measurement systems with JENTEK® MWM™ and IDED sensors to customers in government agencies, as well as those in the private sector. ❖

JENTEK® is a registered trademark of JENTEK Sensors, Inc.
GridStation™ and MWM™ are trademarks of JENTEK Sensors, Inc.

Jentek® Sensors' Gridstation™ Measurement System can provide age degradation monitoring, including fatigue, corrosion, and thermal aging in numerous materials.

Putting the Flex in Flexible

Commercial Benefits—Spinoffs

Image processing and 3-D graphics tools created for the International Space Station serve double-duty by helping Hollywood with special effects, animation, and colorization of old black-and-white television shows and movies.

Dynacs Engineering Company, Inc., is a minority-owned, Small Business Administration-certified engineering company located in Palm Harbor, Florida. In support of NASA programs, Dynacs made advancements in multiprocessor-based computers and software technologies. This work includes symbolic equation processing, graphical user interfaces, and computer animation.

Dynacs Engineering has earned a reputation for its technical capabilities in engineering analysis and modeling of aerospace multibody structural dynamics. Using Dynacs-developed software, a computer model of the Space Shuttle's multi-jointed robotic arm was created. This software-generated model

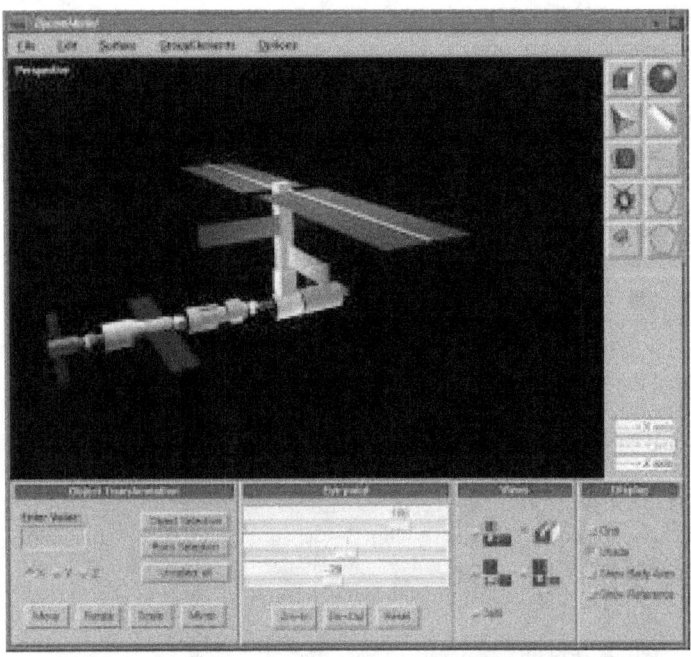

Dynacs Engineering Company, Inc.'s Dycom is a multibody dynamics and control system simulation tool.

proved highly accurate when ground simulations were contrasted with both original manufacturer data and actual data taken during Shuttle missions.

Two Dynacs software products, originally designed for NASA, have been commercialized. GenSoft is a code generation tool. It automatically generates Fortran and C code for mathematical equations. GenSoft was originally used by Johnson Space Center for Space Station simulation. Marketing of the software as a commercial product started in 1993. The product found commercial acceptance in the aerospace world and is in use by prime spacecraft contractors.

A second product, Dycom, was first designed at Marshall Space Flight Center in 1985. Originally called TREETOPS, this computer code was formulated to analyze and design controllers for a number of large space systems, including the Hubble Space Telescope. TREETOPS is capable of simulating the dynamics and control of flexible systems as complex as the Space Station, robot tasks and manipulations, and the rendezvous and docking of spacecraft.

Dynacs was funded to further enhance the software, with the firm acquiring the TREETOPS copyright in 1994. The resulting Dycom software package is now used on such projects as the Space Station, as well as by federal agencies and contractors, including the Defense Department, the U.S. Navy, Lockheed Martin, and the Aerospace Corporation.

Dycom is a multibody dynamics and control system simulation tool. It provides an integrated software environment to perform kinematic and dynamic analysis of space structures and robot manipulators and mechanisms, including their control elements. Dycom users can graphically construct models of mechanical systems, specify forces, torque, and other conditions. The animation package allows zooming into areas of interest. Variations in viewpoint as well as lighting conditions can also be displayed.

Dycom consists of a suite of software tools, all working together in an integrated environment. This powerful product has been made commercially available by Dynacs Engineering. Portions of the software have been adopted by the medical and entertainment industries. In the arena of medical application, Dycom code has been tasked with medical imaging using computed tomography scanners.

NASA has recognized Dynacs for its contributions to space projects on several occasions. In 1997, the firm earned the space agency's Small Disadvantaged Business Prime Contractor of the Year award. The following year, the Johnson Space Center bestowed Dynacs with the Small Disadvantaged Business Contractor of the Year award for the company's contribution to the Space Station team. ❖

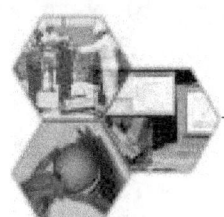

Mathematical Innovation
Commercial Benefits—Spinoffs

Requirements for testing and flight certifying the Space Shuttle Main Engine have spawned a new data analysis and visualization software package.

Through the Small Business Innovation Research (SBIR) program at NASA's Stennis Space Center, MathSoft, Inc., in Seattle, Washington, has developed a system that will provide the building blocks for signal analysis and rapid prototyping. Under the SBIR work, an add-on module to MathSoft's S-PLUS® software has been created. MathSoft develops, markets, and supports technical calculation and data analysis software productivity tools for professionals, students, and educators. The patented S-PLUS® software program is one of the most powerful data analysis software packages presently available.

The SBIR-spurred product, called S+Wavelets®, allows users to perform advanced data visualization and analysis, nonparametric statistical estimation, signal and image compression, signal processing, and to prototype new and faster algorithms. S+Wavelets® can help NASA develop a complete understanding of propulsion test data by using time frequency displays, automatic estimation and de-noising, and data analysis plots for wavelet decomposition. This type of analysis tool is ideal for engineers and scientists and capable of decomposing signals and images into components at different scales and frequencies.

MathSoft was looking for a way to enhance and improve the computational properties, efficiency, and accuracy of mathematical solutions. S+Wavelets® provides analytic and computational properties for this purpose. A number of techniques have been developed to effectively use the unique properties of wavelets. Before S+Wavelets®, scientists and engineers wrote their own code, as there were no commercial wavelet research applications available at that time to support their needs. The Stennis SBIR award provided the resources to create S+Wavelets® and its properties to use waveforms that are localized in time, space, and frequency. In general, the largest benefit of wavelet packets will be seen in the analysis of signals and images with natural oscillations/frequencies. NASA has discovered it is often much easier to remove the noise in the wavelet domain rather than the original domain.

With S+Wavelets®, MathSoft has combined the utilities of the wavelet functions into one comprehensive collection. This has resulted in an efficiency

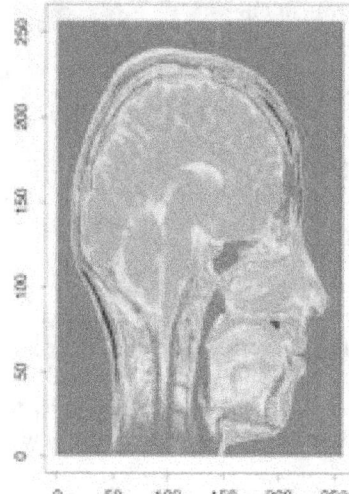
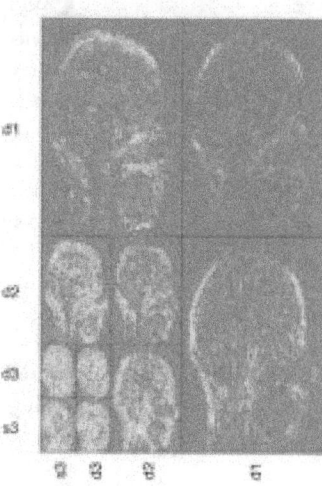

that far exceeds any other mathematical capability known to exist today. The system uses filters to optimize building blocks into which a signal is to be decomposed. These results are used to help interpret the wavelet coefficients and evaluate the decomposition. S+Wavelets® does this by using exploratory signal analysis or rapid prototyping.

As a module of S-PLUS®, S+Wavelets® allows users to develop a thorough, penetrating analysis of signal, time series, or image data. The tool kit offers more than 500 analysis functions within an object-oriented environment. Available for both Unix and MS-Windows platforms, S+Wavelets® is now being commercially sold separately or as an integrated portion of the S-PLUS® software package. Under the same SBIR work, the book *Applied Wavelet Analysis with S-PLUS®* was written.

MathSoft believes the marketing of S+Wavelets® software will soar as wavelet analysis proves to be powerful in easing certain types of computational problems, such as matrix algebra. Various integral and differential equations, when expressed in digital form for a computer, can be solved using matrix algebra.

This mathematical innovation, thanks to NASA's SBIR program, is proving beneficial to scientists and engineers in industry and the government, and is an important software tool for educators and students to solve the most complex of technical problems. ❖

S-PLUS® and S+Wavelets® are registered trademarks of MathSoft, Inc.

MathSoft, Inc.'s software is capable of decomposing signals and images into components at different scales and frequencies.

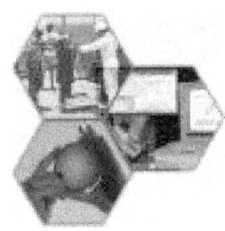

Getting the Lowdown on Airborne Pollutants

Commercial Benefits—Spinoffs

Remote sensing has been taken to new heights given the work of Opto-Knowledge Systems, Inc. (OKSI), of Torrance, California. Through a Small Business Technology Transfer (STTR) award from Goddard Space Flight Center, the company, with the support of NASA's Jet Propulsion Laboratory as the collaborating Research Institute sub contractor, has pushed the frontier of neural network technology for very fast analysis of hyperspectral imagery. The company is also honing the technology of spectral imaging for commercial applications in agriculture, geology, medical diagnosis, manufacturing, and other fields.

The product of a spectral imaging system is a stack of images of the same object or scene, each at a different spectral narrow band, or color. To take full advantage of a hyperspectral system, components must be carefully integrated. Furthermore, special algorithmic tools are needed to analyze and visualize data collected, integrating the entire system into a smoothly functioning instrument.

OKSI has made possible turnkey bench top hyperspectral systems for field or laboratory use. Applications of its hyperspectral camera technology include remote sensing in agriculture, geology, and for military needs, produce and meat inspection, and forensics. Since its establishment in 1991, OKSI has built custom imaging spectrometers in the visible/near infrared and mid-wave infrared portions of the electromagnetic spectrum.

OKSI's Spectral Imaging System is commercially available, with the hardware custom configured for the end-user's needs and requirements. Under Goddard's STTR project, OKSI furthered the advancement in spectral imaging, resulting in several commercial products. Such products include a turnkey bench top hyperspectral imaging system, containing everything the user needs to start applying hyperspectral imaging. The system incorporates the sensor, optics, computer interface, and a versatile hyperspectral analysis software package.

A system sold to a major automobile manufacturer, for example, uses two cameras for stereo imaging and features customized parts and software.

OKSI participated in NASA's Visiting Investigator Program (VIP) under the Commercial Remote

Applications of their hyperspectral camera technology include remote sensing in agriculture, geology, and for military needs, produce and meat inspection, and forensics.

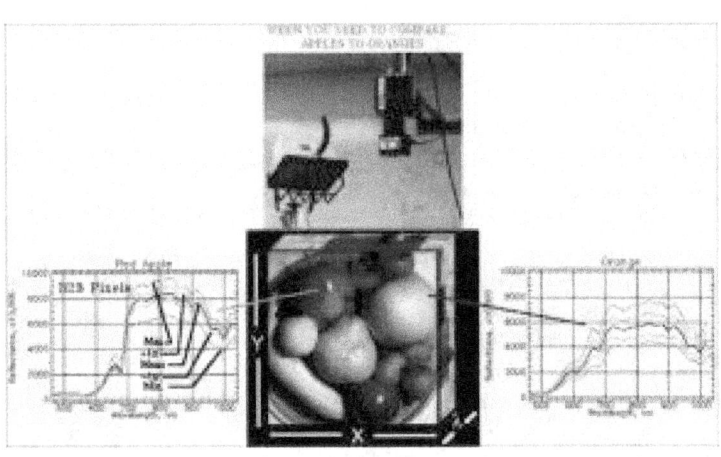

OptoKnowledge Systems, Inc.'s hyperspectral imaging system can be used in applications such as food inspection.

Sensing Program Office at Stennis Space Center. This effort assessed the suitability of thermal infrared imaging data for the remote detection of pollutants and hazardous substances. Stennis remote sensing and OKSI experts worked together to collect data from an aircraft-carried instrument that flew over various areas, including oil refineries and industrial, residential, and agricultural sites. The data were then analyzed, and ground temperature and emissivity maps were generated. Such maps are fundamental requirements for airborne remote detection of pollutants, either in the atmosphere or in the ground, because their spectral signatures are measured against the Earth's background.

As part of this effort, OKSI acquired the spectra of the 189 most hazardous air pollutants as defined by the Clean Air Act Amendment of 1992 and listed by the Environmental Protection Agency. The high-resolution spectra obtained were then processed to demonstrate the capability of a prototype Thermal Infrared Imaging Spectrometer sensor to detect such gases.

OKSI is continuing its work with Stennis. In September 1998, OKSI was selected by Stennis for a 2-year program to demonstrate the use of hyperspectral data in precision agriculture.

This type of research work carried out by OKSI is assisting NASA in developing a suite of airborne and satellite remote sensing applications to study the oceans, biosphere, atmosphere, and land surface. Through the use of algorithmic techniques and data processing, furthered by OKSI's expertise, new ways to assess the health and well-being of the Earth are possible, as are various commercial applications, from resource monitoring to medical diagnosis. ❖

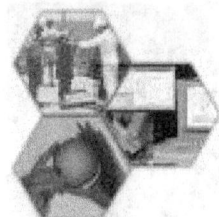

Going Paperless
Commercial Benefits—Spinoffs

You can easily imagine the mounds of paper work required to certify that payloads are ready for a Space Shuttle mission or to approve equipment for installation aboard the International Space Station.

To lessen document overload, NASA-sponsored development of an Electronic Portable Information Collection system, or EPIC for short.

EPIC stemmed from a NASA Small Business Innovation Research (SBIR) project, sponsored by the Kennedy Space Center. During the project, the SENTEL Corporation of Alexandria, Virginia, designed, built, and tested prototype paperless work authorization procedures for NASA. Initially called the Quality Assurance Portable Data Collection system (QAPDC), this project was a key step to the goal of minimizing paper and maximizing quality for Space Shuttle and Space Station operations. In 1996, NASA awarded SENTEL the prestigious SBIR of the Year Award in the computer software category for that work.

QAPDC was a response to manual and paper intensive Work Authorization Documents, record keeping done for all Kennedy procedures. However, this documentation was prone to inconsistency and error. In addition, complexity was increased due to the involvement of several working groups, contractors, and multiple disciplines. Task coordination and verification proved complex and cumbersome.

NASA wanted a system that could convert a procedure from a word processing document to a database. The procedure could then be executed using a portable computer. Data are now entered electronically, either with keyboard or pen, using handwriting recognition. The system then distributes the data to other terminals. Once a procedure is completed, it is electronically stamped and uniquely identified, converted to a Portable Document Format (PDF), and stored electronically in a documentation system. As a prototype of EPIC, QAPDC was tested and shown to be viable. SENTEL and NASA formed a partnership through a nonreimbursable Space Act Agreement. This permitted the go-ahead to develop the operational version of the system, and the EPIC Procedure System was born.

For NASA, the system has proven itself as a rapid and efficient way to collect data, temporarily store data, transfer data to a central personal computer, permit search and retrieval interactions, and generate task status reports and trend analyses. EPIC has been targeted for use in the Space Shuttle Main Engine Shop and for Space Station ground processing.

SENTEL's leadership role in QAPDC and EPIC led to the

Through a partnership with NASA, SENTEL Corporation developed a system that simplifies procedures and documentation.

1998 announcement that the company was teaming with two other firms to commercialize a paperless information collection system. In a major commercial venture, SENTEL has joined with configuration management specialist CMstat and data management industry leader, Symbol Technologies. To provide complete worker mobility, SENTEL is augmenting the information collection system by incorporating Symbol Technologies' Spectrum 24™ wireless networking technology. This partnership is also geared to offer the Advanced Process Manager™.

According to the SENTEL officials, the Advanced Process Manager™ is a state-of-the-art system for executing maintenance and inspection procedures within a robust product data management framework. Primary market targets for the Advanced Process Manager™ exist in the aerospace and defense sectors for life cycle data management. SENTEL believes that marketing of the Advanced Process Manager™ is also appropriate for the airline industries, the shipbuilding industry, warehouse and shipping industries, law enforcement agencies, and public utilities. ❖

Advanced Process Manager™ is a trademark of SENTEL Corporation
Spectrum 24™ is a trademark of Symbol Technologies

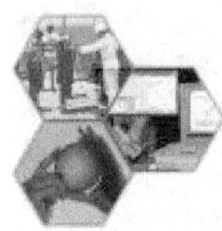

World Reaction to Virtual Space

Commercial Benefits—Spinoffs

While it may be a small world after all, it can be a virtual one as well. DRaW Computing Associates, Inc., of Philadelphia, Pennsylvania, developed a virtual reality software package to support NASA's work on the International Space Station. Through a Small Business Innovation Research (SBIR) project managed by the Marshall Space Flight Center, the company developed software that creates complex virtual reality simulations of the Space Station's numerous components. While the first elements of the Space Station are already circling Earth, astronaut assembly crews face the daunting task of putting together the entire facility in orbit over many years to come. DRaW's powerful software, called OpenWorlds™, is facilitating the training of future assembly teams.

NASA-funded OpenWorlds™ technology is being used for graphics support in virtual worlds on the Internet.

OpenWorlds'™ scripting, hardware, and graphical user interface (GUI) front-end permitted Marshall engineers to "be there without being there," enabling them to fashion the complex habitats, solar arrays, and other central elements in virtual space.

OpenWorlds™ is an open platform for 3-D graphics and virtual reality modeling language (VRML) 2.0 integration. With it, there is the ability to have realistic, interactive, moving worlds. It provides the advanced features of VRML 2.0 without all the effort. The OpenWorlds™ C++ library provides VRML 2.0 support for any application and any graphics server. In addition, sample source code demonstrates implementations of VRML 2.0 browsers on various graphics layers, including OpenGL®. Sample code showing the implementation of the build-in nodes is part of OpenWorlds™, as

OpenWorlds™ is being used for integration of live video and 3-D graphics in next-generation television research.

well as a GUI-based scene graphic viewer. With OpenWorlds™, applications can be made to support Java™ scripting and virtual reality hardware devices.

According to Dr. Paul Diefenbach, president of DRaW, "OpenWorlds™ is an open library accessible by all, open to platforms, open to scripting languages, open to change." VRML is viewed by many as a web-based language for graphics scene description, with additional scripting capabilities throughout. Dr. Diefenbach and his associates viewed VRML differently. Not only is it a powerful tool for graphics exchange, but ultimately, it is an extensible scripting language that also handles graphics scenes. VRML's promise lies in being a medium of exchange for all graphics and simulation systems, Dr. Diefenbach says.

DRaW was incorporated in 1991 and has specialized in 3-D graphics development and consulting, with an emphasis on human factors simulation. DRaW has worked closely with the University of Pennsylvania's Center for Human Modeling and Simulation, the developer of the leading human factors simulation software, Jack®, which is commercially available as Transom™ Jack® from Transom Technologies, Inc.

As a result of NASA SBIR contract awards, OpenWorlds™ has led to the creation of Human OpenWorlds™ (HOW™). Originally an add-on module to Transom™ Jack®, HOW™ provides the power of VRML scripting and transforms human modeling packages into true, interactive, immersive simulation systems by permitting the creation of reactive worlds. Where human factors packages offer control of the human figure, HOW™ provides a realistic virtual world in which the figure can interact. With HOW™, "objects can have behaviors," Dr. Diefenbach says. ❖

OpenWorlds™ is a trademark of DRaW Computing Associates, Inc.
OpenGL® is a registered trademark of Silicon Graphics, Inc.
Java™ is a trademark of Sun Microsystems, Inc.
Transom™ Jack® is a registered trademark of Transom Technologies, Inc.
Human OpenWorlds™ (HOW™) is a trademark of DRaW Computing Associates, Inc.

Software Surrogate
Commercial Benefits—Spinoffs

Since its founding in 1989, providing collaborative software agents has been a core capability developed by Blackboard Technology (BBTech) of Amherst, Massachusetts. As a spinoff of a five-year research project at the University of Massachusetts, BBTech researched, created, and applied cooperative-control technologies to devise software tools capable of interacting between computer programs, thereby mimicking the flexibility of humans, to sort out solutions to problems. Blackboard's proficiency is in building interactive software modules that "brainstorm" as a group to cooperatively solve problems, monitor processes, schedule activities, simulate the future, and perform other complex activities.

Small Business Innovation Research (SBIR) awards from the Johnson Space Center aided BBTech's expertise in the field. NASA requirements centered on the need for human operators to monitor and interact with intelligent computer systems in use aboard piloted space vehicles. BBTech's approach was to develop a generic architecture in which a software surrogate serves as the spacecraft operator's representative in the fast-paced realm of nearly autonomous, intelligent systems. The initial research was carried out under a NASA SBIR award entitled, "A Blackboard-Based Framework for Mixed-Initiative, Crewed-Space-System Applications." This surrogate technology extended the capabilities of BBTech products that are used in a wide range of federal and commercial applications as both a base technology and a commercial software framework.

BBTech became a leader in collaborative-integration applications. This involves closely coupled software systems that work together in varying patterns according to the needs of the application. Some integrated applications pass data from one software system to the next, in much the same way as parts are combined in an assembly line, but this approach is best suited for routine, repetitive tasks. Collaborative-integration, on the other hand, involves using the software systems as a team, to quickly respond to situations.

The firm also enhanced its software products by devising "skeptical agents," tailored to actively monitor the conflicts between organizationally specified activities and potential task-level actions.

BBTech's software has saved users millions of dollars. For example, the Canadian Space Agency reduced its Radarsat-1 mission control system

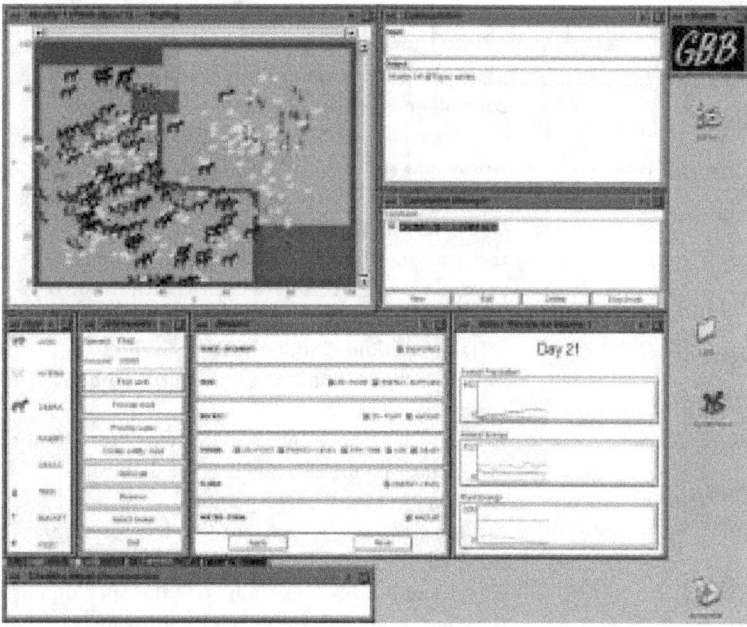

software development costs from an estimated $20 million to $7 million by using the firm's GBB™ (Generic Blackboard Builder) product. By applying GBB™, a satellite scheduling system permits end users anywhere in the world to connect to Radarsat-1's mission control system, to prepare and submit user requests, and to monitor the status and progress of their request as it is planned and executed.

BBTech further developed a line of products, including GBB Runtime™ that provides a low-cost solution for delivering a developed GBB™ application to a large number of users. Another product is ChalkBox™, a graphics toolkit for creating object-oriented, graphical user interfaces that are portable on all hardware platforms on which GBB™ runs.

In 1998, BBTech became part of Knowledge Technologies International (KTI), and portions of past work done under the NASA SBIR program have been applied to KTI's line of products to support knowledge-based organizations. KTI's software products are utilized to create customer-driven product designs for users in aerospace, automotive, and manufacturing industries. The new home for the technology is seen as a strategic step toward the production of innovative, top-of-the-line software. ❖

GBB™, GBB Runtime™, and ChalkBox™ are trademarks of Blackboard Technology.

Blackboard Technology's software "brainstorms" to solve problems, monitor processes, and simulate the future.

Spaceport Support
Commercial Benefits—Spinoffs

The toolkit enables CCT to produce custom command and control systems at commercial off-the-shelf prices.

Having a toolkit at the ready is always wise. But how about for launch vehicles, spacecraft control systems, range control—even entire commercial spaceports?

Located on the Space Coast of Florida, the Command and Control Technologies (CCT) Corporation of Titusville provides high-technology computer system development and other services to the space transportation industry. Over the past 15 years, CCT engineers have been tapped for involvement in both piloted Space Shuttle and expendable rocket operations. More recently, the company assisted in upgrading the Space Shuttle launch control system.

Kennedy Space Center's spacecraft ground processing program is known as the Control Monitor Unit (CMU). Capabilities of this program include processing equipment test data for calibration and diagnosis, controlling the operation of equipment in real-time, simulating the operation of the equipment, and processing large streams of measurement data.

This space center processing technology has been significantly adapted by CCT for use in the commercial sector. The processing program has been licensed and commercialized through CCT. The company is now marketing the technology to the public under the name Command and Control Toolkit™.

CCT and NASA signed a copyright license agreement in late 1997 to commercialize the copyrighted CMU software. The agreement calls for NASA to license the copyrighted CMU software to CCT in return for royalties and other considerations. The company estimates it will recoup the roughly $3 million that NASA has already incurred on design and development of the CMU program. The agreement grants CCT exclusive rights to sell the program to new commercial customers in the U.S. launch vehicle industry.

NASA and CCT are also using the technology for creating new Space Shuttle checkout and launching products and procedures that are being put in place. An advanced portable payload tester prototype, for instance, is undergoing evaluation. Capable of being transported to a payload customer site, the mobile system can reduce processing time and cost at the launch site by readying spacecraft

Command and Control Technologies Corporation's Command and Control Toolkit™ displays such information as engine propellant status.

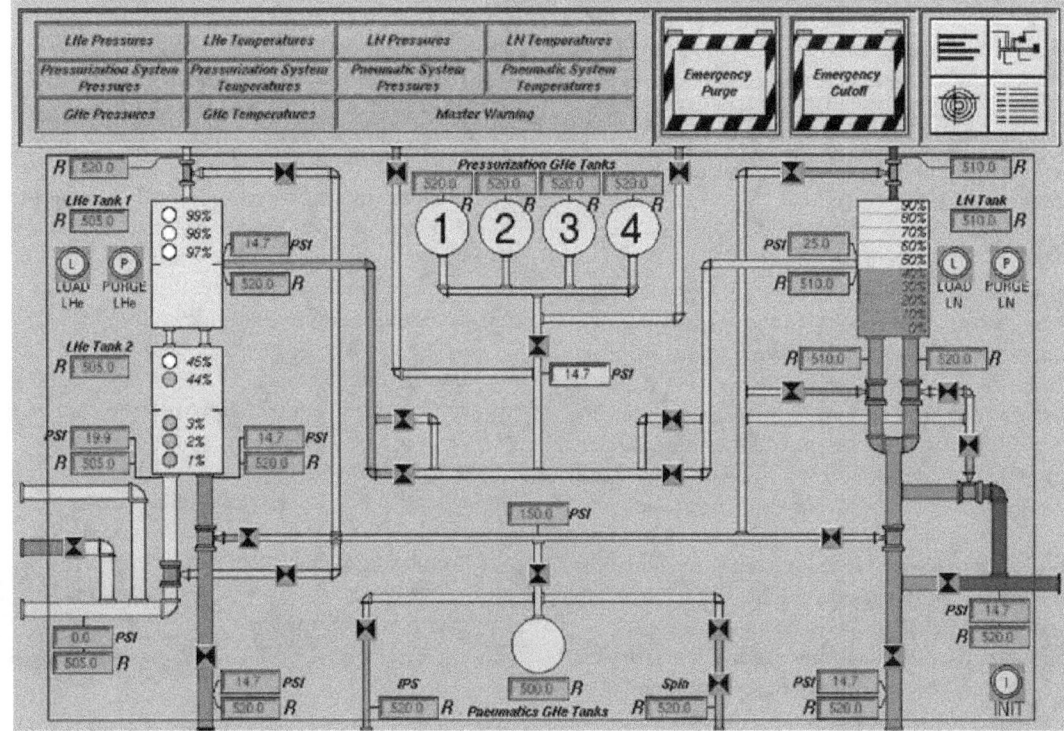

74   Computer Technology

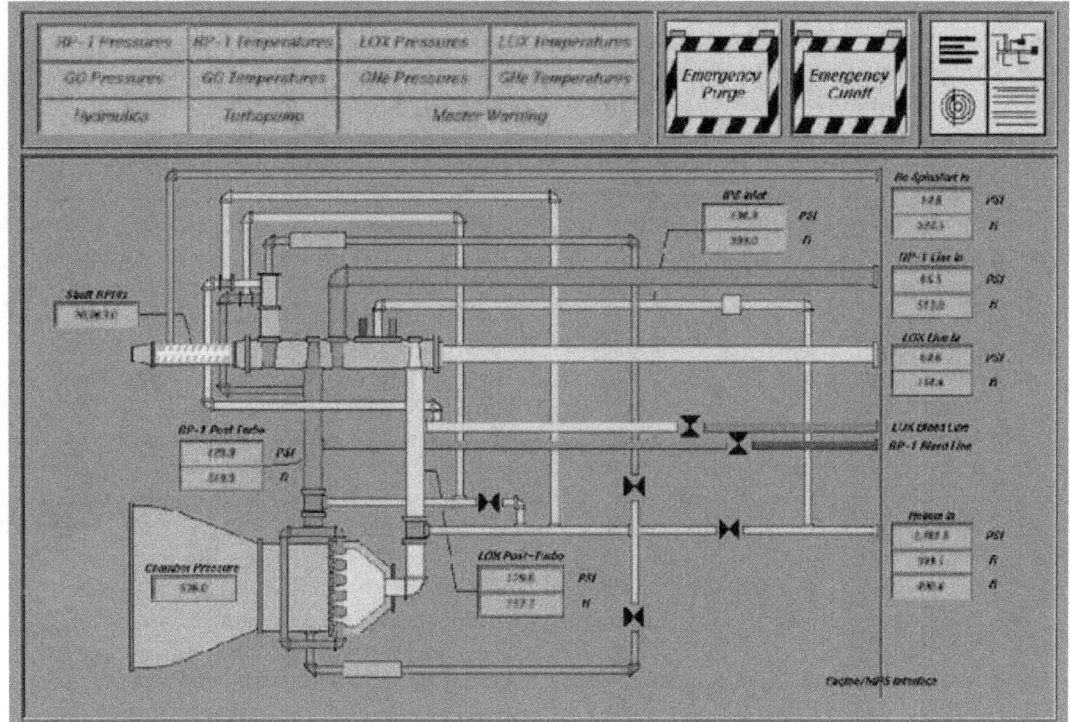

Command and Control Technologies Corporation's Command and Control Toolkit™ displays a variety of information regarding engine status.

more quickly for flight. These virtues are essential for quick turnaround of payloads for 21st century reusable launch vehicles.

The CCT Command and Control Toolkit™ applications are many, including checkout of launch vehicle control systems, range control, ground support equipment control and monitoring, avionics integration and testing, avionics simulations, satellite checkout, and telemetry processing. The toolkit enables CCT to produce custom command and control systems at commercial off-the-shelf prices.

CCT has used the toolkit to design a low-cost spaceport control system for the nation's first operational launch facility, located in southwest Alaska, not owned and operated by the federal government. CCT's toolkit has been used to automate range safety functions at the launch site, along with handling weather data, telemetry and data acquisition, and trajectory planning.

Furthermore, long-term commercial uses for the toolkit are in the offing. These include the remote monitoring of mobile operations, such as offshore oil platforms. Also, remote land operation, using future wireless telecommunications satellite constellates, is among the potential toolkit applications.

In late 1998, NASA selected CCT for the Small Business Subcontractor of the Year Award. The recognition was given to the firm for high technology, real-time software development and computer system design for launch processing and range applications at the Kennedy Space Center. ❖

Command and Control Toolkit™ is a trademark of Command and Control Technologies Corporation

Computer Technology   75

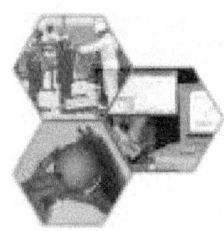

Got an Attitude Problem?

Commercial Benefits—Spinoffs

Satellites circling the Earth demand precision pointing of sensors, antennas, cameras, and other equipment. Without highly accurate angular positioning of spacecraft, a range of dutiful services would not be possible.

An innovative approach to Earth sensor design has been taken by Servo Corporation of America in Westbury, New York. Through a Small Business Innovation Research (SBIR) award from Goddard Space Flight Center, the company built its Mini-Dual Earth Sensor (MiDES). NASA's support for the work stemmed from interest in evaluating a satellite autonavigation system.

The MiDES is an Earth horizon sensor that provides higher accuracy than other types of units, making use of pyroelectric arrays and a digital output. Servo's lightweight, low-cost, dual array horizon sensor was flown in space, confirming its abilities. MiDES is now being eyed for use on a number of commercial telecommunications satellite constellations and other spacecraft to be built both in the U.S. and abroad.

A state-of-the-art Earth horizon sensor, MiDES combines the versatility of a conical scanner with the accuracy and reliability of a staring sensor. The resulting instrument is a lightweight, low-cost, low-power alternative for low-Earth orbiting satellites. Designed specifically to provide horizon attitude position for Earth circling satellites, the MiDES derives its information through the use of two pairs of pyroelectric arrays positioned 90 degrees apart in the imaging plane. Each 16-element array is spatially separated into two 8-element staggered columns that reject and compensate for light emitted by the Sun and reflected by the Moon. The device detects temperature differences between the Earth and space. An algorithm calculates the position of the horizon based on the voltages obtained from the pixels, which subtend the horizon gradient and the pixels that look at space and Earth.

The heart of the MiDES is the hybridized lithium tantalate pyroelectric array. Created by Servo, these lithium tantalate detectors are specially processed, highly sensitive, uncooled detectors that provide very stable outputs over a variety of temperatures and operating conditions. The need for moving parts has been dramatically reduced to near zero. A flex pivot-mounted chopper design eliminates the need for scanning motors, bearings, and lubricants.

MiDES was used on the STS-85 flight of the Space Shuttle Discovery, carried by the Cryogenic Infrared Spectrometers and Telescopes for the Atmosphere (CRISTA)-Shuttle Pallet Satellite II (SPAS). The CRISTA-SPAS was a joint satellite project between NASA and Germany. Once the satellite was deployed from the Shuttle, it flew for 11 days. Several gigabytes of flight data on how MiDES operated were obtained. These data showed that the MiDES accuracy was significantly better than expected, keeping attitude errors well under 0.1 degrees.

Sales of MiDES have begun, with delivery of a unit for flight on a minisatellite program sponsored by the United Kingdom. MiDES is a candidate for several constellates of low-Earth orbiting telecommunications satellites. A variation of the basic MiDES has also been proposed for use aboard several new geosynchronous satellites being designed.

It seems that MiDES has proven itself capable of locking onto new business, as well as providing a solid lock on the Earth's horizon. ❖

Servo Corporation of America's MiDES is a dual array Earth horizon sensor that provides high accuracy for low-Earth orbiting satellites.

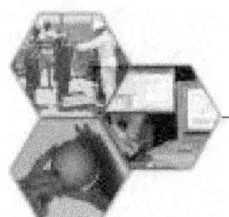

Giving Machines the Vision
Commercial Benefits—Spinoffs

Robots are entering the work force. That might not be so surprising on Earth, but highly capable automatons can also offer a helping robot arm or two to spacewalking construction crews.

Amherst Systems, Inc., of Buffalo, New York, has advanced the visual acuity of robots through contracts sponsored by Johnson Space Center and Langley Research Center. NASA has been evaluating the use of robots as extravehicular companions to spacewalking crews for help and fetch duties.

One such mobile robot developed by Amherst Systems for NASA is the prototype Foveal Extravehicular Activity Helper Retriever, or FEVAHR for short. FEVAHR detects, recognizes, and tracks objects described to it in English, and approaches these objects when instructed to do so, while also dodging obstacles. This robot coordinates its head and body motion to support target search, track, and obstacle avoidance. FEVAHR's visual system is primed by a Semantic Network Processing System so that the robot's vision and behavior are tuned to the task requested.

Central to the robot is the use of foveal machine vision, a new technology that emulates the sophisticated foveal processes found in biological systems. Foveal vision is nature's way of filtering out and reducing the burden of irrelevant information. In fact, a human brain that could process the extraneous information tossed out by foveal vision would have to weigh several tons.

Amherst Systems' approach is to achieve the throughput benefits of foveal vision, but apply it to machinery required to interact intelligently with the real world. The result is a faster, less computer-like response than the response that would have resulted, once all the visual information was processed.

Vision is typically acknowledged as the most "general purpose" method of acquiring information, supporting a wide range of useful tasks (e.g., search, track, recognition, and manipulation). While a traditional camera can collect much information, a majority of this data is irrelevant for any given task.

The charter of Amherst Systems' Machine Vision Department is the development of commercially feasible technology that adopts attractive aspects of biological foveal vision. If robots are to perform useful tasks in a non-deterministic world—rather than being relegated to controlled and very predictable settings like assembly plant floors—they must be imbued with information from which to make decisions.

Amherst Systems' optical and infrared sensor systems have graded acuity and context sensitive sensor gaze control. Imaging hardware is controlled by video processing algorithms. The system can gaze electronically by situating the fovea at different locations within the imager's visual field. The foveal machine vision offers improved performance, greater platform intelligence, and autonomous operation. Lower system cost is attainable, compared to traditional uniform acuity sensors.

This technology is currently being delivered to university research facilities and military sites. Foveal machine vision for mobile robots is intended for applications in poorly controlled environments where response time is critical.

An employee-owned small business, Amherst Systems was founded in 1975. Work in hardware and software development has taken the firm into areas traditionally regarded as separate engineering fields, from sensor chips, pointing mechanisms, and multiprocessors, to visual attention and image processing. ❖

Amherst Systems, Inc.'s FEVAHR detects, recognizes, and tracks objects described to it in English.

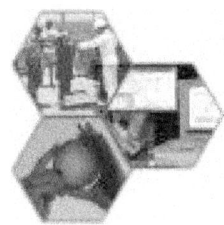

A Particulate Matter
Commercial Benefits—Spinoffs

Even a microparticle of dust or fiber can foul up sensitive equipment being readied for a launch into space. Great care is a must in overseeing the cleanliness of conditions during ground processing of space payloads. A Florida company has found commercial uses for monitoring hardware designed to protect Space Shuttle payloads from mission-compromising contamination.

Technical Applications Unlimited (TAU), Inc., of Cape Canaveral, Florida, received a patent license for the Particle Fallout/Activity Sensor, now called the Real-Time Optical Fallout Monitor (OFM). TAU is a tenant of the Florida/NASA Business Incubation Center in Titusville. Kennedy Space Center's Technology Programs and Commercialization Office arranged this agreement. The firm has commercialized the sensor, developed at Kennedy to detect the accumulation of potentially damaging dust and fibers on delicate payload components.

TAU is marketing the OFM, highlighting the product's attributes for use in clean rooms, semiconductor manufacturing, pharmaceutical production, spacecraft processing, and food processing. Further applications include the monitoring of medical facilities, security systems, and air handling assessment.

Invented by NASA engineers, the sensor measures relative amounts of dust or other particles that have fallen onto a special wafer. Using an infrared light-emitting diode (LED), an optical assembly, and other microelectronics, fallout accumulations can be measured. The prototype sensor included a sensor module. The commercial OFM comes as a single unit. TAU improvements have also enhanced the instrument's sensitivity.

The OFM is portable, compact, and weighs little more than three pounds. The unit can detect single particles as tiny as 10 microns. Fallout accumulation measurements can be displayed on the OFM's front panel and/or downloaded for remote monitoring analysis.

The role of the Florida/NASA Business Incubation Center, established in 1996, is to assist entrepreneurs and small businesses by offering office space at a reduced cost and providing technical assistance to tenants. The center is located in Titusville, situated on the campus of Brevard Community College.

A computer chip maker purchased OFM units to assess the ability of the hardware to detect sources of contamination that can lead to defective chips coming off the assembly line. An aerospace company purchased a unit for use during integration of its spacecraft, the first of which was a Titan IV rocket at the end of 1998. A bio-pharmaceutical manufacturer is evaluating the unit too, for inclusion in their manufacturing process to spot environmental intrusions that render their products unusable. A major nuclear laboratory is considering the OFM as a possible aid to detect environmental fallout in various experimental nuclear processes. Also, a major U.S. photographic firm is eyeing the OFM as an aid to assess cleanliness in film manufacturing.

TAU president John Horan, Jr., explains, "NASA has made it possible for TAU to get off the ground with a high technology product that would have taken years and many more resources than a typical small business start-up company can make available. We feel like TAU will be able to compete with the instrumentation industry leaders as a result of this boost from NASA."

"We have several units now successfully sold to, and in current use by, high technology companies who have environmental monitoring needs that are consistent with the instrument's capabilities. Those companies range from spacecraft manufacturers to integrated circuit manufacturers," Horan adds. ❖

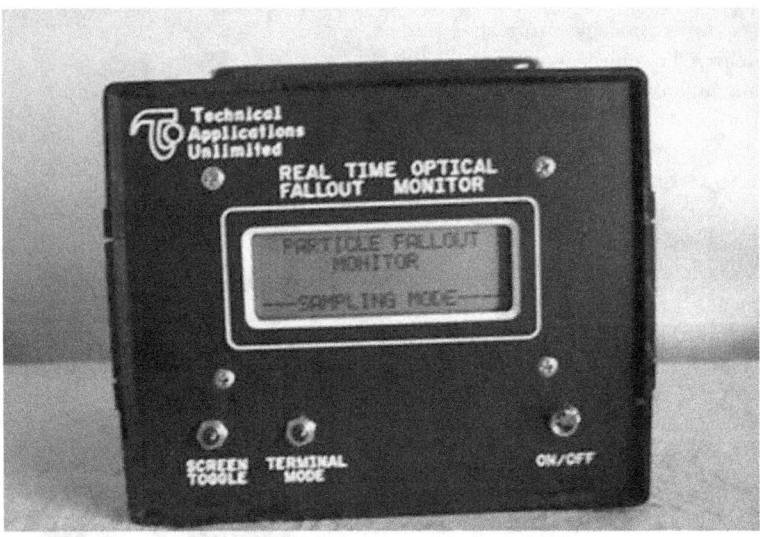

Technical Applications Unlimited, Inc.'s Real-Time Optical Fallout Monitor is designed for use in many areas of industry.

Paint and Shoot
Commercial Benefits—Spinoffs

Up Polariscope! That's the call if you need to monitor such items as fatigue in composite materials or visualize residual strains in glass and plastic.

A few years ago, through a Small Business Innovation Research (SBIR) award from NASA's Langley Research Center, Stress Photonics, Inc., of Madison, Wisconsin, developed the Delta Therm 1000. This infrared-based stress measurement system was commercialized, permitting its user to compare designs and detect cracks in structures, especially aging aircraft and bridges. Combining digital signal processing technology with a special infrared camera, the system provided instantaneous thermal images and live differential images.

Stress Photonics has furthered its research on structural integrity analysis. Under a Langley Small Business Technology Transfer (STTR) contract, the firm has perfected the portable Grey-field Polariscope. This device, called the GFP 1000, involves a single rotating optical element and a digital camera for acquiring full-field images quickly and efficiently.

The company has also introduced specially tinted brush-on coatings. While paint-on coatings have been possible for many years, they were rarely used because the thickness of the coating directly affected the stress measurement.

For the GFP 1000 to efficiently measure the coating thickness, a coating tint attenuates red, green, and blue light to different degrees. Once a target object is illuminated with the GFP 1000's circularly polarized light, oscillations of that light can be reviewed for strain amplitude and direction. When measured by the Polariscope, for instance, a paint-on coating applied to a wrench would reveal the tool undergoing shear strains as a bolt was being tightened.

Moreover, by using the GFP 1000, the thickness of a brush-on coating can be ascertained. Therefore, the Polariscope that measures strain amplitude can also appraise the thickness of the coating, eliminating it as a variable. The result is a full-field strain map by using a truly simple "paint and shoot" technique. Data are displayed on a laptop computer for analysis. Data collected are presented in an intuitive format so that they are easy to interpret. One image depicts the shear strain magnitude with gradients represented in grayscale levels, and the other indicates the directions of the principal strains. Commercially available, the GFP 1000 also works with all currently available

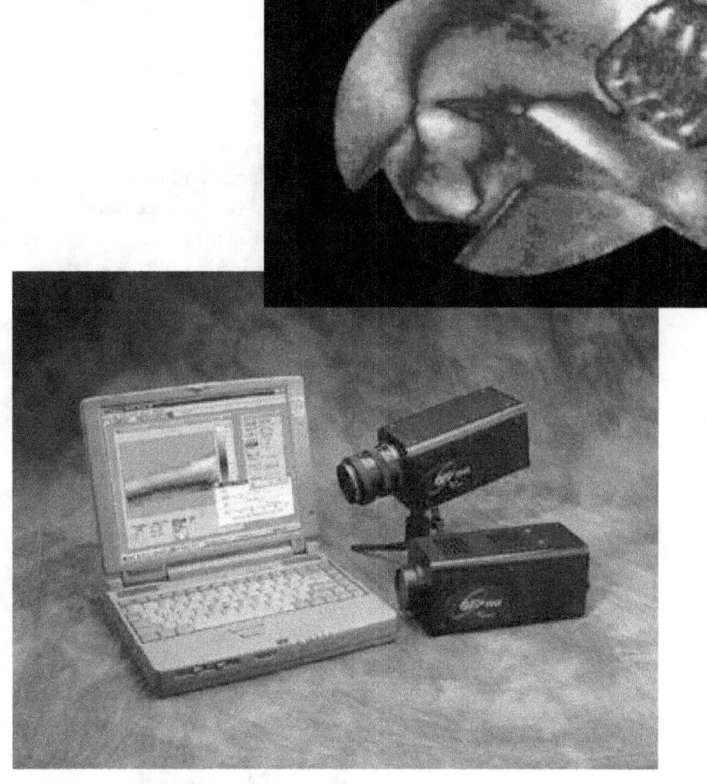

Stress Photonics, Inc.'s GFP 1000 acquires full-field images quickly and efficiently, detecting cracks in structures.

photostress coatings, such as paste-on sheet or liquid contoured.

Reflective photoelastic strain analysis has a longstanding history. Although proven to be very effective over decades, the technique has been largely overlooked by scientists as a fruitful area for innovation due to its maturity. Supported by NASA STTR funds and enabled by Stress Photonics researchers, this tried and true technique now finds greater usefulness by incorporating modern video and computer technologies.

More than $2 million in research and product development projects have been completed by Stress Photonics. The company is one of Wisconsin's top SBIR recipients, receiving research and development funds from several U.S. government agencies, as well as major industries. It actively markets a growing line of stress and strain analysis and flaw detection products worldwide. In 1994, the organization was selected by more than 75 leading scientists to receive the prestigious R&D 100 Award. ❖

Portable Inspection
Commercial Benefits—Spinoffs

The harsh environment of space can be extremely unforgiving on satellite exteriors, thermal control surfaces, and power-providing solar cells. A hand-held device has been added to a spacewalking astronaut's toolkit, one that permits easy and reliable inspection of spacecraft surfaces. Such equipment is critical in determining maintenance schedules for thermal control surfaces of large and complex spacecraft, including the International Space Station. Work on the portable instrument for space has led to a variety of commercial products that are ideal for industrial applications.

AZ Technology, Inc., of Huntsville, Alabama, developed the Space Portable Spectro Reflectometer (SPSR) under NASA's Small Business Innovation Research (SBIR) program. In partnership with the Marshall Space Flight Center, the company designed the lightweight and portable instrument for use aboard the Space Shuttle. The company also created the Optical Properties Monitor experiment that was conducted on the exterior of Russia's Mir space station.

Drawing upon this NASA-sponsored work, AZ Technology has produced several commercial instruments, and has several more in development. One such device is the TESA 2000 (Total Emittance and Solar Absorptance), developed in partnership with NASA's Kennedy Space Center. The TESA 2000 portable inspection instrument is a powerful combination of two instruments, packaged for ease of use in the field or laboratory. As the name suggests, the apparatus measures both the near room temperature emittance and solar absorption of a material in order to fully characterize its radiative heat transfer properties. The product is designed to pack into two carrying cases that easily fit into the overhead bin of most airliners. Rugged and highly accurate, the unit is designed for years of field use and includes two rechargeable batteries and a battery charger.

Evaluations by NASA of the hardware included ground-processing tasks of the Shuttle and its payloads within the cargo bay. Specifically, the TESA 2000 determined the effectiveness of thermal insulation blankets and radiator surfaces. These blankets protect sensitive instruments from the ultra-heat and cold, as well as from damaging doses of space radiation. The tests suggested that surface inspection costs of a Space Shuttle could be cut in half, thanks to the equipment. Furthermore, it proved less bulky than older equipment utilized, thus the device was safer to use, particularly in tight, hard-to-access areas.

Commercial applications of the TESA 2000 include evaluating special coatings used by military on land, sea, and air vehicles. Coating manufacturers and testing laboratories could use the technology to measure surface reflectance properties of coating panels in environmental tests for weathering.

Additionally, TESA 2000 can evaluate the efficiency of solar cell or solar heating systems, and coatings and reflectance materials utilized in automotive glass can be assessed by manufacturers or testing laboratories.

AZ Technology has been able to develop a range of instruments through a vigorous technology transfer program with NASA. The company has also made space-stable, electrically conductive thermal control coatings resulting from NASA SBIR awards. Coupled with the company's expertise in materials testing and development of advanced spacecraft coating materials, the firm offers "one stop shopping" for spacecraft coating needs, as well as the ability to address many commercial needs. ❖

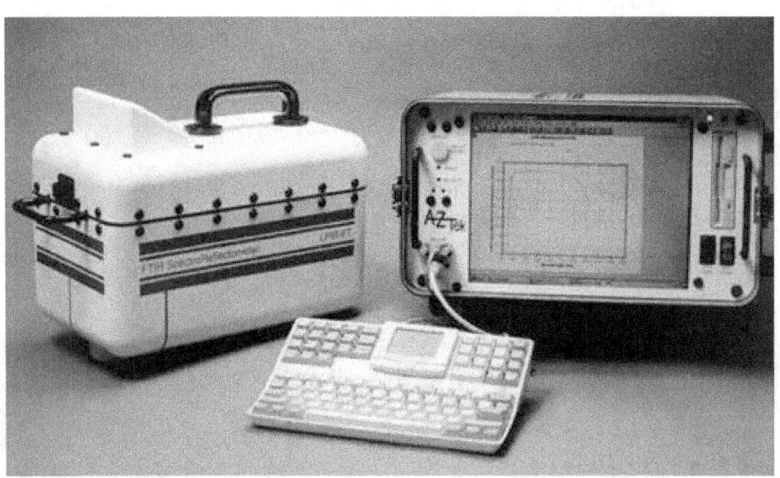

AZ Technology, Inc.'s TESA 2000 is a portable inspection instrument, packaged for easy use in the field or laboratory.

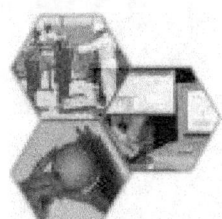

Ultra-Precision Optics
Commercial Benefits—Spinoffs

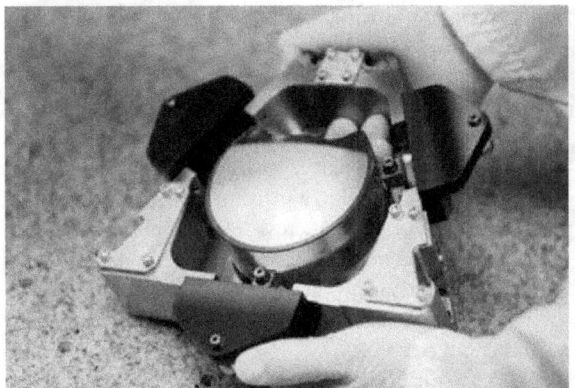
Courtesy of LLNL and EUV, LLC

Extreme ultraviolet (EUV) light enables greater accuracy than current lithography systems.

Precision optics are essential to orbiting space telescopes as they reveal the outer reaches of the surrounding Universe. Demand for those tolerances is having an impact on down-to-earth lithography technology used in producing semiconductors.

A collaborative effort has been established that meshes the talents of NASA's Goddard Space Flight Center with leaders in semiconductor manufacturing. By leveraging U.S. government and semiconductor industry resources, the program is expected to yield unprecedented advancements in reflective optics for the semiconductor industry.

The collaborative effort involves Goddard, SEMATECH, Silicon Valley Group Lithography (a division of Silicon Valley Group), and SVG-Tinsley. This four-party research and development initiative—dubbed the High Precision Optics Joint Sponsored Research Agreement (JSRA)—is to harness the wherewithal to create the most advanced microlithography machines to date. Lithography is a process that makes possible the printing of tiny electronic circuitry, such as those used in powerful computer memory chips. Higher precision optics of benefit both to NASA and for advancing optics used in lithography are being realized by the JSRA. The JSRA also involves the pooling of resources to reduce costly research and development expenditures.

An early decision was made to use extreme ultraviolet (EUV) light for a prototype lithography optics design. Tolerances for EUV optics form accuracy requirements that are typically five times more stringent than for lithography systems currently in production.

The JSRA was stimulated, in part, by the Richmond, California-based SVG-Tinsley's successful fabrication of ultra-precise corrective optics for the space agency's Hubble Space Telescope that was deployed in 1990. Astronomers were at first dismayed when they found the space telescope's primary mirror flawed. That problem greatly impaired the optical vision of the orbiting eye-in-the-sky. To correct Hubble's focus, SVG-Tinsley was assigned by NASA the critical task of making sets of coin-sized mirrors exhibiting extremely smooth surfaces. These corrective optics permitted Hubble's full and fabulous telescopic power to be realized.

In the government-industry collaboration, Silicon Valley Group Lithography (SVGL) of Wilton, Connecticut, plays the role of systems integrator. SVGL will apply newly developed ultra-precision technologies to future microlithography products. This higher precision enables lithography systems to yield breakthrough optical performance, resulting in higher productivity, improved overlay, increased process latitude, and a larger exposure field.

Goddard brings to the collaboration ultraviolet mirror coatings, special test facilities, analytical modeling, and other technical know-how. SVG is a leading manufacturer of automated wafer processing equipment for the worldwide semiconductor industry. The company designs, manufactures, and markets technically sophisticated equipment used in the primary steps of making semiconductors. SEMATECH is a consortium of U.S. semiconductor manufacturers. SEMATECH works with government and academia to sponsor and conduct research aimed at assuring the country's leadership in this high-technology arena.

This collaborative powerhouse is expected to benefit state-of-the-art technologies, be it in microlithography or scanning the heavens above in never-before-seen detail. "Continuing improvements in optics manufacturing will meet unique NASA requirements and the production needs of the lithography industry for many years to come," notes Noreen Harned, SVG's director of corporate lithography marketing. ❖

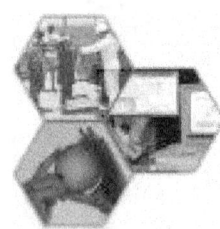

Rapid Prototyping
Commercial Benefits—Spinoffs

Javelin's unique SteamRoller™ System provides a rapid prototyping process for making ceramic components.

The ability to use ceramic materials for space-based components has helped reduce product development cycles in equipment manufacturing via rapid prototyping.

Accomplishments under NASA Small Business Innovation Research (SBIR) and Small Business Technology Transfer (STTR) contracts have helped enable Javelin, a division of Lone Peak Engineering, Inc., of Salt Lake City, Utah, to become a dominant developer in the advancement of rapid prototyping technologies. The SBIR and STTR efforts were conducted for NASA's Johnson Space Center and Marshall Space Flight Center, respectively.

Using the most advanced systems available today, Javelin has designed one of their most notable developments, the SteamRoller™ System, a rapid prototyping process that produces functional ceramic prototypes directly from computer-generated models. This fully automated ceramic-laminated object manufacturing—CerLAM™ for short—is a process that builds ceramic prototypes layer-by-layer using ceramic sheets. The prototype is built directly from a computer-generated model without any tooling.

Rapid prototyping is an exploding new technology that is evolving within the design and manufacturing industries. This technology makes it possible to fabricate physical parts, without the need for tooling, through the direct conversion of three-dimensional data. Rapid prototyping uses the principles of slicing, layering, and bonding to build the part.

Three-dimensional data is first sliced into cross-sectional planes by a computer. These planes are sent from the computer to the rapid prototyping machine, which builds the part layer-by-layer. The first layer of the part is bonded to a platform or starting base. The shape of the first cross-sectional plane defines the part's geometry. The second layer is bonded to the first and shaped according to the second cross-sectional plane. This process is repeated until the part is complete.

Javelin's SteamRoller™ System couples the CerLAM™ build process to the conventional laminated object manufacturing (LOM) machine. Prior to the SteamRoller™, the build process was semi-automatic and required an operator to position each ceramic sheet in the LOM machine for lamination and cutting. Recently introduced by Javelin, the commercially available SteamRoller™ System allows both ceramic and metal sheet materials to be rapidly prototyped on the LOM.

Engineered ceramic materials that have been used by Javelin's SteamRoller™-LOM process include: 99.9 percent alumina, zirconia, silicon carbide, aluminum nitride, silicon nitride, aluminum silicates, hydroxyapatite, and various titanates.

The CerLAM™ process gives design and manufacturing engineers a rapid, cost-effective, and flexible option for ceramic prototyping or when small quantities of parts are needed. A companion to the CerLAM™ process has been developed at Javelin for metal prototypes and tooling as well.

Initially, Lone Peak Engineering, under which Javelin operates, conducted contract research studies. However, the company's focus has shifted to emphasize design, prototyping, and production of industrial components out of engineered models, such as technical ceramics and advanced polymers. Several developmental efforts have been pursued, such as producing erosion resistant parts, corrosion thwarting pumps, laser guides, and automotive seals.

The company has also manufactured the following products: ceramic golf shoe spikes that never flatten; zirconia ceramic-bladed razor blades for microtomes that never rust, corrode, or need sharpening; ceramic turbines with a hardness approaching that of diamonds and a strength equaling the best German steels; and even bioceramic bones prototyped directly from medical data. ❖

SteamRoller™ and CerLAM™ are trademarks of Lone Peak Engineering, Inc.

A Force to Reckon With

Commercial Benefits—Spinoffs

From the development of smaller heart pumps, more compact audio speakers, robotic "bugs" that walk, and controls for airflow in automobile engines, to quieting the roar of aircraft engines, society can benefit from these advancements and more—all due to a little wafer with a big "wiggle." NASA's piezoelectric wafer technology has now been commercialized, sparking a broad range of potential applications in industry and scientific marketplaces.

NASA's Langley Research Center invented and patented the Thin-Layer Composite-Unimorph Ferroelectric Driver and Sensor (originally called THUNDER®, a trademark later registered by Face International Corporation). This technology is also known as Prestressed Piezoelectric Composites (PPC).

Several years ago, NASA researchers were exploring the well-known phenomenon exhibited by piezoelectric materials, which generate mechanical movement when subjected to a voltage. Such a property can be applied in electronics, optics, noise cancellation, pumps, valves, suppression of irregular motion, and a variety of other fields. This technology can also be used as a sensor in such applications as microphones, non-destructive testing, and vibration testing.

A remarkable feature of these devices is their ability to provide inordinately large mechanical output displacements, as high as 40 to 50 times the thickness of the device itself. That "wiggle" is an order of magnitude greater than existing devices operating in the same frequency range. What's more, these composite piezoelectric structures are tougher than current commercially available piezoelectric materials. The revolutionary devices have greater mechanical load capacity and can easily be produced at a relatively low cost, lending themselves well to mass production. The fabrication process for these devices is readily controllable, resulting in a highly uniform production.

NASA has granted licenses to two Virginia-based companies: Face International Corporation of Norfolk, Virginia, and Virginia Power and Electric Company of Richmond, Virginia.

Face International has successfully commercialized its line of THUNDER® piezoelectric wafers. While offering piezoelectric actuators and sensors as standard products, Face International also sells "made-to-order" wafers, integrating customers' special configurations, hence, fulfilling the custom needs of clients.

Face International has exclusive license to develop actuator systems suitable for shaking concrete and processing other slurried materials. The company owns patents for using controlled acoustic energy (i.e., vibration or sound) to achieve the rapid setting of freshly poured concrete. They are also developing THUNDER®-based tools for surface finishing of wet concrete. These tools will impart certain vibrations to plastic concrete, facilitating the finishing of the slab. According to Face International officials, using piezoelectric devices to generate "smart" vibrational energy to work fresh concrete has never been done before. In addition, the company own patents for THUNDER®-based pumps, switches, and circuit breakers.

Face International is selling a variety of THUNDER® devices and is now capable of producing these and similar devices by the thousands on a monthly basis. By the first quarter of 2000, the company's manufacturing capacity is forecast to extend into the tens-of-thousands per month. Plans call for fully automated high-speed, high-volume production that can churn out quantities of a million-plus by the end of 2000.

Development of Virginia Power's NASDRIV™ devices is currently underway. The company is authorized to sublicense in a wide range of applications, excluding concrete-related applications.

The commercial possibilities for this small and forceful technology are staggering. ❖

THUNDER® is a registered trademark of Face International Corporation
NASDRIV™ is a trademark of Virginia Power and Electric Company

Face International's THUNDER® devices provide inordinately large mechanical output displacements.

Optical Fiber Protection

Commercial Benefits—Spinoffs

The NASA Small Business Innovation Research (SBIR) Program provided the needed research funds, as well as technology, for the development of a technique to protect sensitive strands of optical fiber from harsh environments. This technique has helped fashion products useful in taking strain and temperature measurements, as well as in spotting trace amounts of biological and chemical warfare agents. NASA funds and expertise went into helping F&S, Inc., (soon to be named Luna Innovation) of Blacksburg, Virginia, develop ruggedized coatings and coating technologies that are applied in the making of optical fiber.

NASA's requirements prompted F&S to develop a metal coating capability specifically for space-based applications of optical fiber sensors. The unique and demanding performance required from the space-rated optical fiber led to a novel metal coating technique developed by F&S: on-line sputtering of very thin metal coatings onto optical fiber during production.

The F&S optical fiber fabrication facility and various coating practices have enabled the company to make specialty optical fiber with custom designed refractive index profiles and protective or active coatings. The company has demonstrated sputtered coatings using metals and ceramics, and combinations of both. Also, the company has developed ways to apply thin coatings of specialized polyimides formulated at Langley Research Center. With these capabilities, F&S has produced cost-effective, reliable instrumentation and sensors able to withstand temperatures of up to 800 degrees centigrade. Yet another outcome of its research is the demonstration of sapphire fiber-based sensors capable of performing at temperatures above 1500 degrees centigrade.

F&S has adapted the same sensing platforms to provide the rapid detection and identification of chemical and biological agents with parts-per-trillion sensitivities. More sensitive than devices now used, the firm's fiber optic "biosensor" also does the job far more quickly.

F&S is continuing work with NASA through additional contracts with the Dryden Flight Research Center, as well as Langley. The company is developing flight-worthy systems for physical measurements through Dryden, and work is underway with Langley to develop novel materials in microcomposites, high displacement actuators, and active polymers.

F&S innovations in self-assembled monolayer thin film coatings are being realized. This is a capability with a wide variety of applications. Lightweight, actuated composite structures, for instance, will allow deployment and surface control of large spaceborne and terrestrial optical and electronic receivers. Such equipment could find application for astronomy and communications purposes, at a very reduced cost for fabrication, maintenance, and operation.

F&S has patented optical fiber fabrication techniques, devices, and instrumentation for the physical measurement of strain, temperature, and pressure, as well as chemical and biological detection, explains Michael Gunther, company vice president and co-founder. F&S has also designed and fabricated microelectromechanical systems (MEMS) products. MEMS devices include optical switches, micropumps, and optical interconnects.

Gunther explains that since the company was founded in 1990, worldwide product sales have continued to grow annually. The firm is dedicated to the transition of basic research to cost-effective commercial products. ❖

NASA's requirements prompted F&S to develop a novel metal coating technique.

Advanced Polymers for Practical Use

Commercial Benefits—Spinoffs

The need for improved, more cost effective satellite and land-based electronic applications has prompted the development of a high performance, low-cost substrate for printed circuits. This advancement in liquid crystal polymers (LCPs) is also ideal for high barrier packaging for foods and beverages, high performance tubing, barrier layers for cryogenics, and high temperature electrical insulation.

The technology for multilayer printed circuit boards was developed under a Small Business Innovation Research (SBIR) contract between NASA's Goddard Space Flight Center and the polymer division of Foster-Miller, Inc., of Waltham, Massachusetts. The commercial viability that resulted from the original research was so substantial that Foster-Miller created a spinoff company, Superex Polymer, Inc., also of Waltham, Massachusetts.

LCPs have a propensity to self-align during the liquid state, forming oriented regions that can be further constructed into fiber, film, or molded parts. Because of this alignment capacity, once an LCP layer is formed, its structure persists, becoming resistant to such stresses as melting heat. LCP thin layers can be incorporated into blends that offer more favorable performance and cost-competitive advantages over other materials.

Because of Superex's efforts in processing LCPs, exceptionally strong, stiff, and lightweight tubing is being manufactured for endoscopic instruments. The tubing was the winner of the prestigious R&D 100 Award and has been licensed to a medical equipment company.

LCPs can also be used as a "super barrier" packaging material. This is partly due to their oxygen transmission level—six to eight times higher than ethylene vinyl alcohol under humid conditions. By using LCPs as thin high barrier layers, their properties can prevent oxygen from deteriorating the taste of precooked and packaged food. Also, the LCP water vapor transmission level exceeds many fluoropolymers. Additionally, LCP films have excellent barrier properties for carbon dioxide, nitrogen, and other gases. Compared with polymers typically used for packaging, the price of LCP resin is high, but its proportionately higher properties give LCPs favorable performance tradeoffs, according to Rick Lusignea, president of Superex Polymer, Inc.

New processing and packaging technology is advancing, such that layering of various films, along with LCP, is possible to create bottles, trays, and jars. Packaging applications targeted for LCPs include: multilayer films for all-plastic retort pouches and bag-in-boxes with long shelf-lives; multilayer food trays with a shelf-life of one year; LCP barrier layers for disposable medical bags; LCP laminates for use in high-barrier bags for industrial chemicals; plastic liners for reusable steel tanks; and LCP-based plastic lids for snack food cups and trays to replace metal lids.

Superex is pioneering blow-molding machinery, bolstering the company's prospects for moving into a variety of key markets.

There are still many other possible applications for this technology, including the use of LCP laminates in compact, high-speed computers, robust high-density circuit boards, multilayer boards, microwave and high-speed digital circuit boards, multichip modules, and flexible printed circuits. ❖

Superex Polymer, Inc.'s liquid crystal polymers can be used in the production of bottles.

Superex Polymer, Inc.'s liquid crystal polymers can be used to produce exceptionally strong, stiff, and lightweight tubing for such uses as endoscopic instruments.

Wearout Free

Commercial Benefits—Spinoffs

Cornell Dubilier Electronics, of Wayne, New Jersey, has designed, developed, and is selling two types of capacitors as a direct result of Small Business Innovation Research (SBIR) contracts with the Marshall Space Flight Center.

NASA demands were for a space-rated, lightweight, high-energy density capacitor. The goal was to exceed the performance attained with wet tantalum capacitors widely used in space applications.

A key requirement for this kind of electronic part is handling the release of hydrogen gas. Employing a double seal with material to absorb the hydrogen in an aluminum electrolytic capacitor is one approach, but proves very expensive to produce due to the complex assembly required for the double seal, in addition to the capacitor not being hermetic. The double seal design made use of a rubber grommet seal at the bottom of the seal chamber and a glass-to-metal seal at the top.

Losing electrolytes through evaporation limits a capacitor's usefulness in military and space applications requiring long life. Thanks to the use of nongassing electrolytes, manufacturing hermetic aluminum electrolytic capacitors that exhibit no "wearout" from electrolytic depletion became possible. Based on the SBIR work, Cornell Dubilier created an aluminum electrolytic capacitor, one that is hermetically sealed, incorporating weld-seal technology and low gassing electrolytes.

Cornell Dubilier's spinoff capacitor featured the welded closure and low gassing electrolyte of the NASA capacitor, but it did not have the outer glass-to-metal seal. Even so, this resulted in a capacitor with a very tight seal, eliminating 95 percent of the rubber seal area of conventional capacitors. The rate of electrolyte loss from this new capacitor was one-hundredth the rate of loss in a similarly rated, conventional aluminum electrolytic capacitor.

Life tests and measurements of weight loss at high temperatures showed expected operating lifetimes for the spinoff capacitor of 50 to 100 years. That is well beyond the 10 years expected of typical aluminum electrolytic capacitors.

Another advancement from the NASA-supported work was reshaping the cylindrical capacitor by

Life tests and measured weight loss at high temperatures showed expected operating lifetimes for the spinoff capacitor of 50 to 100 years.

Cornell Dubilier created a low gassing electrolytic capacitor.

taking advantage of welded-closure technology. Corners are not possible with conventional capacitors because they produce poor seals with the required rubber gasket materials. A flatpack shape was developed, one looking more like a pocket-carried lighter with leads and mounting tabs.

Electrolytic capacitors are well known for their use in a variety of electronic equipment used for data processing, communication, entertainment, and power electronics.

Cornell Dubilier has brought to market MLP and MLS Flatpack Aluminum Electrolytic Capacitors. Dubbed the MLP for "military, low profile," the spinoff flatpack capacitor found its first application in modules used for a military power supply system. The original NASA capacitor has evolved to become quite useful in applications where high capacitance is needed in a box shape.

The MLP's high-energy storage and box shape lends well to power-hold-up modules, with MLPs stacked together and sealed in metal boxes. Such modules are used for power-hold-up in F-16 fighter aircraft. Another application benefiting from the MLP's flatpack shape is a new ground-to-satellite phone system in which the MLP is a battery stiffening capacitor. The MLP is in the hand-held phone and delivers the high-peak currents needed to transmit to satellites in orbit. ❖

Out of the Lab...Into the Real World

Commercial Benefits—Spinoffs

No longer are lasers confined to laboratory bench tops. Today, lasers are finding their niche in the fields of medicine, imaging, ranging, restoration, remote sensing, and in measuring our environment. As more and more users take lasers out of the lab, however, these powerful tools must perform reliably in "real-world" situations. Often large, delicate, and complicated to use, scientific lasers don't fit the bill for a variety of needs.

Big Sky Laser Technologies (BSLT), Inc., of Bozeman, Montana, is the developer of small, rugged commercial and developmental laser systems. Company engineers have addressed three key features demanded by laser users: equipment that is compact, rugged, and turnkey. A capacity for reliable laser operations in almost any environment was the chief design goal of the company. As a result, BSLT has introduced its Compact Folded Resonator (CFR) line-up of lasers. The CFR series is a family of highly reliable Nd:YAG lasers, crafted to take on a wide array of duties.

Through the help of TechLink of Bozeman, an organization that works to support the commercialization of NASA and other federal laboratory technologies, BSLT contacted NASA. The company's engineers determined that Langley Research Center's patent for detection and control of prelasing in a Q-switched laser would be of value to their line of products. Prelasing is a condition that occurs when light "leaks" out of the laser cavity prematurely. By applying a way to detect and control prelasing in a Q-switched laser, the deleterious effects of prelasing can be avoided.

Subsequently, BSLT applied for a non-exclusive license of the NASA technology, employing it within the company's CFR 800 laser unit. TechLink is continuing to assist the firm in commercial development and applications of the NASA technology after licensing.

As the latest addition to its CFR line of compact, rugged laser sources—the CFR 200 and CFR 400—the powerful CFR 800 features a laser head over eight times smaller than comparable systems. The CFR 800 boasts a quick and easy lamp change for its single lamp design and a total head weight of just 25 pounds. The CFR 800 is energized by a microprocessor-controlled power supply for maximum user flexibility and cooled by a closed loop, de-ionized water-to-air heat exchanger.

With laser heads up to 90 percent smaller and 75 percent lighter than scientific lasers, the CFR series allows users to take lasers where they could not go before. Taking a scientific laser out in the field can be a risky proposition. Dust, condensation, or other foreign matter inside the laser head can spell trouble. To counter these real-world worries, the CFR series of lasers is sealed with O-rings. Hermetically sealed, the laser systems can operate reliably in even the most unfavorable environments.

Applications for the CFR 800 are numerous and span a wide range of possibilities, such as Light Detection and Ranging (LIDAR), remote sensing, eye-safe illumination, ablation, and marking. The CFR 800 is marketed by BSLT as perfect for many applications in industry, science, manufacturing, research and development, and military use. Due to its small size and high-energy output, interest in the CFR 800 for a variety of uses is on the increase, according to the company. ❖

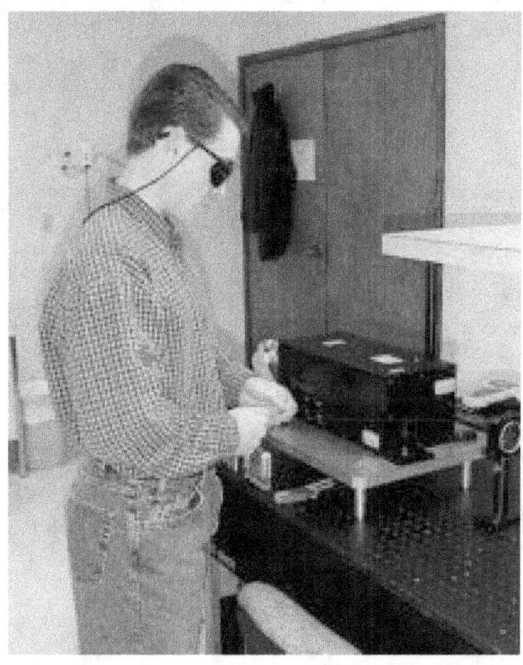

Big Sky Laser Technologies, Inc., developed a product line of small and rugged lasers that not only can be used in the lab, but also in "real-world" situations.

Illuminating Development

Commercial Benefits—Spinoffs

How best to keep an eye on high-temperature processes where extreme brightness might otherwise obscure the view?

Initiated by support from Small Business Innovation Research (SBIR) funds from the Glenn Research Center, Control Vision, Inc., of Idaho Falls, Idaho, has commercialized a series of laser-augmented, video sensor technologies.

Control Vision's first product line, LaserStrobe®, was developed under a NASA SBIR effort over 12 years ago. This novel equipment yields clear, high-resolution, real-time video imaging of high-temperature, high-energy industrial processes. Welding, plasma arc spraying, arc furnaces, metal casting, and refractories melting are among the processes that LaserStrobe® can monitor. The Control Vision systems use reflected laser or strobe illumination, combined with ultra-short double exposure techniques, that allow for particle imaging velocimetry of fast-moving powder particles buried within a plasma stream.

Basically, with LaserStrobe®, the laser pulse creates the video image while ignoring the brightness coming from the process. The sensor is essentially

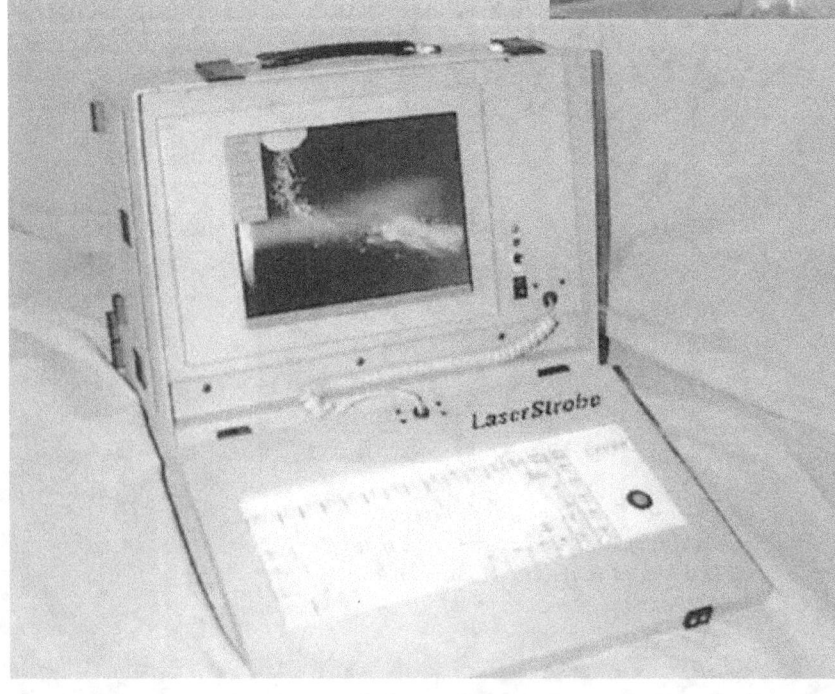

Control Vision, Inc.'s LaserStrobe® and PyroCam® systems can yield high-resolution video images of high-energy industrial processes.

88 Industrial Productivity/Manufacturing Technology

blind to the radiation coming from the process. The rapid-fire laser pulse also freezes powder particles in flight.

The LaserStrobe® is well suited for conventional welding, laser welding, and other laser-driven processes, as well as thermal spray, metallurgy, and ceramics research.

A few years ago, the company introduced the PyroCam® line of imaging systems. PyroCam® technology was originally developed for Nd:YAG laser welding processes and was derived from research and development efforts also funded under the NASA SBIR program.

The PyroCam® system includes a small sensor head, a xenon strobe illumination unit, and a microprocessor-based controller unit. Since its introduction, Control Vision has found motion analysis applications for PyroCam®, such as observation of machining operations. Additionally, high-temperature applications, such as viewing ceramics within a high-temperature furnace, have been made feasible by PyroCam®. The system works well for peeking in on processes where temperatures can be as high as 2000 degrees Centigrade. PyroCam® has been used in steel mills to observe processes like melting, continuous casting, hot rolling, and cutting.

NASA's interest in PyroCam® has recently increased, specifically to tap into the equipment's attributes that allow more cost-effective in-process weld inspection of aluminum-lithium or other alloy materials to be used in the Space Shuttle's Lightweight External Tank. The system is expected to assist NASA in the quality and cost efficiency of Space Shuttle External Tank assembly work at the space agency's Michoud plant in New Orleans, Louisiana.

Under a recent contract with NASA, Control Vision has developed a new version of the LaserStrobe® system. This newer device, the Model 4Z, replaces the conventional controls of its earlier counterpart with all-digital, computerized controls. Model 4Z's optics package provides users a variety of magnifications and close-up optics with varying standoff distances to accommodate research needs. Each of the two compact pulsed nitrogen lasers includes a fiber optic cable that is used to deliver the laser illumination to the area being watched. This technology has significantly added to the theoretical and scientific knowledge of plasma diagnostics and plasma processing.

Control Vision's leadership in high resolution, real-time imaging of high temperature processes is proving itself in both private sector and governmental research and development projects. ❖

LaserStrobe® and PyroCam® are registered trademarks of Control Vision, Inc.

NASA Success and Education—A Special Feature

With the 21st century close at hand, maintaining the competitiveness of American industry is paramount. Technology is a major ingredient in retaining a national leadership role within the world community. The great storehouse of technological know-how that the National Aeronautics and Space Administration (NASA) embodies can foster applications far different than those for which they were originally intended, as these pages of *Spinoff* clearly demonstrate.

But along with NASA-developed expertise, there are additional programs that are immensely valuable, and integral to the production of the treasure trove of space technologies and secondary applications.

NASA has established numerous technology outreach programs, offering technical assistance to support small start-up companies to large enterprises, as well as city and state organizations. NASA's skills are being transferred from the space agency's own research laboratories to the public domain to help solve challenging concerns, such as strategies to reduce ozone levels in the home, joining efforts to clean water in polluted bays, or to help save babies following corrective fetal surgery.

Promoting excellence in education is another area in which NASA is active. Each of the four NASA Strategic Enterprises—Aero-Space Technology, Earth Science, and Human Exploration and Development of Space, Space Science, — are steeped in inspiration, wonderment, and the electricity of gaining new knowledge. Through an array of initiatives, NASA is devoted to supporting teachers and enthralling the students of today. This educational mission is focused on increasing the public's awareness of how science, mathematics, and technology offer the foundation to help enrich the quality of life in the new millennium.

Outreach Achievements

NASA Success and Education—A Special Feature

Problem solving is what NASA has long been about. The space agency's men and women offer their talents, time, and facilities to work out difficult challenges that confront our nation. Each year, there are many success stories that underscore NASA's problem solving abilities and a willingness to cultivate partnerships for the public good.

One such helping-hand effort is assisting the Smithsonian Institution in its three-year project to help give new life to "Old Glory" by preserving the Star-Spangled Banner. An infrared camera built at Goddard Space Flight Center for exploration of Mars is in use to image the historic flag.

The Mars camera's infrared abilities are well suited to help preservationists identify deteriorated and soiled areas on the flag that are not obvious to the human eye. The flag is suffering from decades of exposure to light, air pollution, and temperature fluctuations, despite special care given the historic icon at the Smithsonian's National Museum of American History. Things that are difficult to see, or invisible to the human eye, can be detected by scanning the flag in infrared. Moisture and dirt, for instance, are culprits that can degrade the flag. Goddard's Acousto-Optic Imaging Spectrometer (AimS), and its special ability to make an image with reflected infrared light, is perfect for the job. Goddard's camera team is also exploring the use of AimS in skin cancer research, as well as using it to study the pigment used in paints, which can help distinguish an authentic piece of art from a forgery.

A Sensors 2000 team at the Ames Research Center has produced a "pill transmitter," which monitors body temperature, pressure, and other vital signs in the womb, and radios this critical information to physicians. Nearly every time doctors operate on a fetus, the mother will later undergo pre-term labor that must be monitored. Pre-term labor is a serious problem that is difficult to predict and monitor with conventional equipment, and often leads to the death of the baby. Thanks to the Ames effort, prototype versions of pill-shaped devices are being designed. Placed in a woman's body through endoscopic surgery techniques, these devices can measure and transmit data on acidity in the fetus, with the future of measuring electrical activity of the fetal heart. These pill transmitters are also being studied for monitoring physiological changes in astronauts during space travel.

As increasing numbers of aircraft crowd the skies, NASA and the Federal Aviation Administration (FAA) have joined forces to improve safety and air traffic. The partnership is geared toward improving aviation safety, airspace system efficiency, and aircraft environmental concerns. The agreement to work together guarantees that NASA's abilities and the FAA's air transport industry expertise will be combined to provide a safer aviation system and an affordable and dependable service for the air traveler and the air carrier business.

For over 20 years, a foam developed by researchers at the Ames Research Center has been used in many applications. From wheelchairs to airplane passenger seats, the material, often referred to as temper foam, is one of the most widely used NASA inventions. The material is a cell polymeric foam with "slow springback" properties.

Researchers at the Marshall Space Flight Center have been using a NASA aircraft to fly over U.S. cities, like Baton Rouge in Louisiana, Sacramento, California, and Salt Lake City, Utah. Supported by NASA's Earth Science enterprise, science teams are flying a thermal camera to take each city's temperature profile. Images taken by the camera can pinpoint the cities' "hot spots." These are bubble-like accumulations of hot air, called urban heat islands. The bubbles of hot air develop over cities as naturally vegetated surfaces are replaced with heat-radiating asphalt, concrete, rooftops, and other human-made materials.

The Marshall science team is working with the trio of cities initially surveyed to incorporate results into the cities' plans. This type of research is expected to promote strategic planning, such as where best to plant trees and use light-colored, reflective building material, to help cities maintain their cool. NASA's environmental work can help urban planners reduce ozone levels, focus tree-planting programs, and best locate heat emitting roofs.

NASA has also pledged to help in water cleanup and pollution-reduction efforts in the Chesapeake Bay and surrounding areas. The space agency and the Environmental Protection Agency (EPA) have entered into an agreement to better understand the Chesapeake Bay ecosystem and its impact on surrounding communities. NASA is contributing a wealth of data it has gleaned from satellite sensors, research aircraft, and the Space Shuttle to help the environmental cause. This information includes water temperature measurements and data on pollution runoff, algae bloom, and fish populations. The NASA/EPA agreement covers new data, as well as historical measurements from the space agency's remote sensing archives. Armed with this type of information, environmental officials can establish priorities for cleaning up the bay and outlying areas. The bay watershed is extensive, covering New York, Pennsylvania, Maryland, Delaware, Virginia, West Virginia, and the District of Columbia.

Yet another technology assistance from NASA involves outreach to industry in identifying natural marine oil seeps in the Gulf of Mexico, offering clues on oil deposits. Through the Commercial Remote Sensing Program at NASA's Stennis Space Center, NASA is demonstrating practical applications of space technologies in America's marketplace.

Oil migrates naturally through cracks from deposits deep below the ocean floor, releasing into the world's surface waters. These marine oil seeps offer clues as to where oil deposits may be located in ocean basins. Marine oil seeps occur naturally and are manifested as oil slicks on the ocean's surface.

Through Stennis, industry partnerships are being established to use remote sensing and related technologies to explore markets that can enhance opportunities for industry customers. NASA's technical and financial participation helps reduce the market risk associated with new product development to a level that partnering companies can accept. By way of Stennis' outreach to industry, small companies can explore the use of remote sensing without exposure to excessive financial risk.

Be it saving a historic flag, helping reshape the hot and bothersome landscape of the city, or creating life-saving technologies for babies, NASA know-how is being leveraged in many ways to sustain and enhance the quality of life on Earth. ❖

The Educational Frontier

NASA Success and Education—A Special Feature

Hundreds of elementary, secondary, and higher education activities and programs are being supported by NASA and its field centers. A brief sampling of the vitality of NASA's obligation to inspire students and teachers through learning opportunities includes the following activities.

Clearly, the power of the Internet continues to grow as more than a library, a television, or a pathway to passive information. One of its most exciting applications is as a tool for collaboration. It allows teachers and students to work together to design inquiries and explore challenges.

Managed at the Ames Research Center, NASA's Quest Team offers projects built around interactions between students and working space professionals. Current projects include Space Scientists Online, which explores such hot topics as astrobiology—the study of life in the universe. Live events on the Internet (both chats and audio/video programs) are offered on a diverse set of topics within NASA's Office of Space Science. These projects—called Sharing NASA—permit students to share the excitement of space agency pursuits like flying the Space Shuttle, living aboard the International Space Station, exploring distant worlds with spacecraft, and pushing the frontiers of aeronautical research.

At the Jet Propulsion Laboratory (JPL), partnerships have been put in place with the Boys & Girls Clubs of America, International Technology Education Association (ITEA), the University of Southern California, and the Astronomical Society of the Pacific (ASP). JPL's web page, "The Space Place," is an exciting new Internet site designed primarily to introduce young students and their teachers to some of the latest and most advanced technologies for use on future space missions. Specifically, JPL's New Millennium series of spacecraft are highlighted, whose missions include probing beneath the surface of Mars or voyaging to the heart of a comet. Internet users are offered a combination of educational activities and related technical and scientific fun facts.

Barbara Morgan, an elementary school teacher from Idaho, is being trained at the Johnson Space Center to take education to new heights. The tragic Challenger accident in 1986 claimed the lives of the Space Shuttle crew, including teacher Christa McAuliffe. Over 12 years later, McAuliffe's successor has entered a rigorous schedule for space travel aboard a future Shuttle mission. Morgan will become the first educator mission specialist. Morgan has spent more than a decade working with NASA on space education projects and programs. She is preparing to engage young minds in subjects like math and science, making space real for teachers and students around the world as she and her crewmates circuit the Earth in a Space Shuttle.

Computers are tools of the future. In just one year, NASA has donated over 36,000 excess computer items with an original cost of $75 million to public, private, and parochial schools serving students in pre-kindergarten through the 12th grade. NASA is taking an active role in the federal Computers for Learning program, a White House inspired activity that streamlines the transfer of excess federal computer equipment to those U.S. schools with the greatest need.

A special partnership was established in 1998 between the Dryden Flight Research Center and National History Day, an effort funded by the National Endowment for the Humanities, as well as other foundations, corporations, and individuals. Select teachers from California attended a workshop to learn more about the history of space technology. Approximately 20 teachers from across the state toured Dryden , reviewed the center's history, and familiarized themselves with the site's educational resources. Also, new lesson plans were gathered to assist their students in developing projects for the National History Day program, with the theme of Science, Technology, and Invention in History.

Space-enthusiastic students and teachers across the country took an active part in observing the flight of John Glenn aboard Space Shuttle Discovery in October 1998. NASA centers were engaged in school programs from sixth graders at Harrisburg Christian Academy in Pennsylvania to Ohio's Muskingum College, Glenn's alma mater. During Glenn's flight aboard Discovery, students at the John H. Glenn High School in New Concord, Ohio, the Center of Science and Industry (COSI) in Columbus, Ohio, and the Newseum in Arlington, Virginia, participated in audio-only education events, communicating with Glenn as he flew through space.

A stellar opportunity presented itself as engineers at the Marshall Space Flight Center teamed with students and faculty at Brigham Young University (BYU) in Provo, Utah. The objective of the collaboration was to test an inexpensive telescope viewing x-ray emissions from Earth's neighboring star, the Sun. NASA and university teams are readying a fully automated, x-ray telescope to be used aboard the Space Shuttle, a project stemming from a BYU student's research paper in 1987. GoldHelox is the project's name, inspired by the golden color of the Sun and "helox" standing for HELiocentric Observations in X-rays. The entire educational project will only cost about $200,000 for its 13-year lifetime.

In celebration of NASA's 40th anniversary, Bandai America, creators of the popular Power Rangers action figure series, launched a special line of action figures called "Heroes of Space." The likenesses of Apollo astronauts have been fashioned, packaged with one of the popular Power Rangers in Space action figures. Each assortment comes with a mini-comic book on the history of the Apollo flight, a trading card, and replicas of accessories used by the moonwalking astronauts in their missions, such as a rock box, American flag, camera, and space helmet.

In late 1998, the NASA Office of Equal Opportunity Programs selected nearly a dozen minority universities to receive a Precollege Award for Excellence in Mathematics, Science, Engineering, and Technology (PACE/MSET) grant. Nine other minority universities received a Minority University Mathematics, Science and Technology Awards for Teacher and Curriculum Enhancement Program (MASTAP) grant.

Each PACE/MSET university receives up to $100,000 per year for the three years of the grant, based on performance and availability of funds under the program. The PACE grants provide opportunities for minority colleges and universities in collaboration with NASA and local school districts. The key objective is to provide for informal educational opportunities that will enhance the number of students enrolled in college preparatory courses in mathematics and science. These partnerships are designed to encourage more students to enroll in college mathematics, science, engineering, and technology (MSET) disciplines and to pursue MSET careers in the future.

The MASTAP grants provide opportunities for minority colleges and universities to develop diverse and exemplary research-based mathematics, science, technology, and geography teacher education curricula, integrated with content from NASA's mission. Each MASTAP grant award recipient receives up to $200,000 per year for the three years of the grant based on performance and availability of funds under the program.

Through a series of events in 2003, NASA, the Department of Transportation, and the Federal Aviation Administration (FAA) have teamed to commemorate the 100th anniversary of the Wright brothers' first powered flight. To highlight the 95th anniversary of that historic aviation event in 1903, special aviation education events were held that introduced students to transportation careers and the high-tech skills required to attain air and space transportation goals in the 21st century. More than 700,000 students and 10,000 teachers nationwide became a part of the special programming carried over the NASA CONNECT instructional television series focusing on the anniversary program. Viewers of the Public Broadcasting Service were also able to experience the series. ❖

NASA's Commercial Technology Network

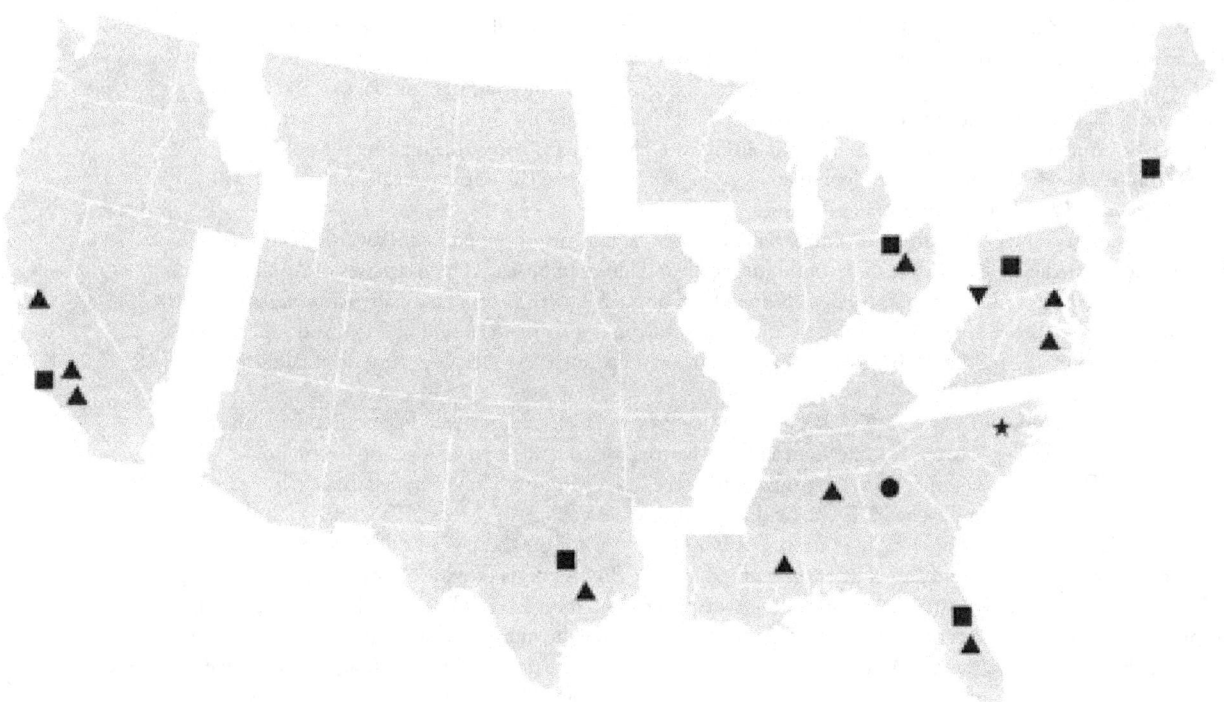

The NASA Commercial Technology Network (NCTN) extends from coast to coast. For specific information concerning commercial technology activities described below, contact the appropriate personnel at the facilities listed or go to the Internet at <http://nctn.hq.nasa.gov>. General inquiries may be forwarded to the National Technology Transfer Center.

To publish your success about a product/service you may have commercialized using NASA technology, assistance, or know-how, contact the NASA Center for AeroSpace Information or go to the Internet at:

<http://www.sti.nasa.gov/tto/contributor.html>.

▲ **Field Center Commercial Technology Offices**
Represent NASA's technology sources and manage center participation in technology transfer activities.

▼ **National Technology Transfer Center**
Provides national information, referral, and commercialization services for NASA and other government laboratories.

■ **Regional Technology Transfer Centers**
Provide rapid access to information, as well as technical and commercialization services.

★ **Application Team**
Provides a range of technology management services including technology assessment, valuation and marketing; market analysis; intellectual property audits; commercialization planning; and the development of partnerships.

▲ FIELD CENTERS

Ames Research Center
National Aeronautics and
   Space Administration
Moffett Field, California 94035
Chief (acting), Commercial Technology Office:
Carolina Blake
Phone: (650) 604-0893
email: cblake@mail.arc.nasa.gov

Dryden Flight Research Center
National Aeronautics and
   Space Administration
Post Office Box 273
Edwards, California 93523-0273
Chief, Public Affairs, Commercialization and
   Education Office:
Lee Duke
Phone: (805) 258-3802
email: duke@louie.dfrc.nasa.gov

Glenn Research Center
National Aeronautics and
   Space Administration
21000 Brookpark Road
Cleveland, Ohio 44135
Chief, Commercial Technology Office:
Larry Viterna
Phone: (216) 433-5398
email: larry.a.viterna@grc.nasa.gov

Goddard Space Flight Center
National Aeronautics and
   Space Administration
Greenbelt, Maryland 20771
Chief, Commercial Technology Office:
George Alcorn, Ph.D.
Phone: (301) 286-5810
email: galcorn@pop700.gsfc.nasa.gov

Lyndon B. Johnson Space Center
National Aeronautics and
   Space Administration
Houston, Texas 77058
Director, Technology Transfer and
Commercialization Office:
Henry Davis
Phone: (281) 483-0474
email: henry.l.davis@jsc.nasa.gov

John F. Kennedy Space Center
National Aeronautics and
   Space Administration
Kennedy Space Center, Florida 32899
Associate Director, Technology Programs and
Commercialization Office:
James A. Aliberti
Phone: (407) 867-6224
email: jim.aliberti-1@ksc.nasa.gov

Langley Research Center
National Aeronautics and
   Space Administration
Hampton, Virginia 23681-0001
Director, Technology Applications Group:
Samuel A. Morello
Phone: (757) 864-6005
email: s.a.morello@larc.nasa.gov

George C. Marshall Space Flight Center
National Aeronautics and
   Space Administration
Marshall Space Flight Center, Alabama 35812
Director, Technology Transfer Office:
Sally A. Little
Phone: (256) 544-4266
email: sally.a.little@msfc.nasa.gov

John C. Stennis Space Center
National Aeronautics and
   Space Administration
Stennis Space Center, Mississippi 39529
Manager, Technology Transfer Office:
Kirk V. Sharp
Phone: (228) 688-1914
email: kirk.sharp@ssc.nasa.gov

Jet Propulsion Laboratory
4800 Oak Grove Drive
Pasadena, California 91109
Manager, Commercial Technology Program Office:
Merle McKenzie
Phone: (818) 354-2577
email: merle.mckenzie@jpl.nasa.gov

▼ NATIONAL TECHNOLOGY TRANSFER CENTER

Wheeling Jesuit University
Wheeling, West Virginia 26003
Joseph Allen, president
Phone: (304) 243-2455
email: jallen@nttc.ed

■ REGIONAL TECHNOLOGY TRANSFER CENTERS

1-800-472-6785
You will be connected to the RTTC in your geographical region.

Far-West
Technology Transfer Center
University of Southern California
3716 South Hope Street, Suite 200
Los Angeles, California 90007-4344
Kenneth E. Dozier, Jr., director
Phone: (213) 743-2353
email: kdozier@bcf.usc.edu

Mid-Atlantic
University of Pittsburgh
3400 Forbes Avenue, 5th Floor
Pittsburgh, Pennsylvania 15260
Lani Hummel, executive director
Phone: (412) 383-2500
email: lhummel@mtac.pitt.edu

Mid-Continent
Texas Engineering Extension Service
Texas A&M University System
301 Tarrow Street
College Station, Texas 77840-7896
Gary Sera, director
Phone: (409) 845-8762
email: ecsera@teexnet.tamu.edu

Mid-West
Great Lakes Industrial Technology Center
25000 Great Northern Corp. Ctr., Suite 260
Cleveland, Ohio 44070-5320
Christopher Coburn, executive director
Phone: (216) 734-0094
email: coburnc@battelle.org

Northeast
Center for Technology Commercialization, Inc.
1400 Computer Drive
Westborough, Massachusetts 01581
William Gasko, Ph.D., director
Phone: (508) 870-0042
email: wgasko@ctc.org

Southeast
Southern Technology Application Center
University of Florida
College of Engineering
1900 S.W. 34th Street, Suite 206
Gainesville, Florida 32608-1260
J. Ronald Thornton, director
Phone: (352) 294-7822
email: jrthornton@ufl.edu

★ TECHNOLOGY APPLICATION TEAM

Research Triangle Institute
Post Office Box 12184
Research Triangle Park, North Carolina 27709
Doris Rouse, Ph.D., director
Phone: (919) 541-6980
email: rouse@rti.org

NASA CENTER FOR AEROSPACE INFORMATION

Spinoff Project Office
NASA Center for AeroSpace Information
7121 Standard Drive
Hanover, Maryland 21076-1320
Jutta Schmidt, Manager
Phone: (301) 621-0182
email: jschmidt@sti.nasa.gov

Pictured left to right: Zoe Rush, *editor;* Deborah Drumheller, *publications specialist;* John Jones, *graphic designer;* Jutta Schmidt, *manager;* David Ferrera, *editor;* Walter Heiland, *consultant.*

Not pictured: Leonard David, *writer;* Kevin Wilson, *photographer.*

www.ingramcontent.com/pod-product-compliance
Lightning Source LLC
Chambersburg PA
CBHW081731170526
45167CB00009B/3783